AF762271

5.50

Dychweler y llyfr h[illegible] neu
pan elwir amdan[illegible]
Rh[illegible]

Book Runners Order PO NYAATP08

ORDER N

PO NYA

BOOK N

9907

BAY NU

144

AUTHOR	Draper, S. R.
TITLE	Biochemical Ana
DESCRIPTION	

UMBER

ATP08

UMBER

940

MBER

)-11

lysis in Crop Science

Biochemical analysis in crop science

Biochemical analysis in crop science

SIMON R. DRAPER
DEPARTMENT OF AGRICULTURE
UNIVERSITY OF ABERDEEN

UNIV. COLL N.W.
BANGOR
LIBRARY

OXFORD UNIVERSITY PRESS
1976

771972

Oxford University Press, Walton Street, Oxford OX2 6DP
GLASGOW NEW YORK TORONTO MELBOURNE WELLINGTON
CAPE TOWN IBADAN NAIROBI DAR ES SALAAM LUSAKA ADDIS ABABA
DELHI BOMBAY CALUTTA MADRAS KARACHI DACCA
KUALA LUMPUR SINGAPORE HONG KONG TOKYO

CASEBOUND ISBN 0 19 854128 7
PAPERBACK ISBN 0 19 854132 5

© OXFORD UNIVERSITY PRESS 1976

All rights reserved. No part of this publication may be reproduced, stored in a retrieval system, or transmitted, in any form or by any means, electronic, mechanical, photocopying, recording or otherwise, without the prior permission of Oxford University Press

SCIENCE
QK 865.O73

FILMSET IN HONG KONG
By T.P. Graphic Arts Services
PRINTED IN GREAT BRITAIN
By Fletcher & Son Ltd., Norwich

Foreword

This book will interest many people in plant biochemistry as an extremely relevant science for securing progress in crop production for the future. The importance of such progress is obvious at a time when parts of the Third World are already seriously short of food, when many more parts have increasingly grave difficulties in the face of population explosions, and when in both Western and centrally controlled economies the importance of reliable sources of food and feeding stuffs has, once again, been made dramatically clear.

Many young scientists, striving to master recent explosions of knowledge, find it hard to turn their appreciation of basic concepts and related theory to practical purposes, particularly in such broad and complex subjects as crop production and utilization. They have great difficulty in finding brief, readable accounts of the intricate results already obtained, and of the practical methods of obtaining them. They need a reliable guide to purposes and techniques related to the taxonomy of molecular types of proven importance. They also need careful selections of references so that particular matters can be pursued in greater depth as purposes and interests may later require. And many with years of experience of applied biochemistry will also find value in such a wide-ranging guide to give them a fresh view of developments outside their own particular specialities, and perhaps even to change their area of specialization.

Dr. Draper has planned his useful book so as to attract and help both those who specialized early in biochemistry and those whose first interests were in other sciences and their integrations in agriculture and horticulture. He has produced a comprehensive, but brief and readable, guide which also has a full bibliography, making it a useful reference book.

PROFESSOR J.R. RAEBURN, C.B.E.
Strathcona-Fordyce Professor of Agriculture,
The University of Aberdeen.

Acknowledgements

DURING the preparation of the manuscript the author was greatly assisted by many colleagues in the School of Agriculture, Aberdeen. Special thanks are due to Professor J.R. Raeburn for writing the Foreword and for making many helpful suggestions and to Dr. T.B. Miller for advice and comment on many detailed points. The help of Drs. H.F. Walker, J.K. Thompson, and A.M. Innes is also gratefully acknowledged. In addition, the author wishes to thank Mr. W. Sandison for skilled technical assistance with some of the experiments mentioned in the text, Mrs. S. Gill and Mrs. E. Watt for preparing the typescript, and Miss H. Kenley for drawing the figures.

Contents

Note

At the end of Chapters 3–6 of this book the references are arranged in two lists, under the headings *Books and review articles* and *Research papers*. They are referred to in the text by the name of the author (and the date of publication).

1. Introduction

THE efficient production of food by the cultivation of crop species is the essential basis of intensive agriculture. The crop scientist has already played a central role in many of the major developments which have allowed our present systems of farming practice to evolve, and it is unlikely that further improvements will be achieved without a continued scientific analysis of the factors limiting crop production. To the crop scientist involved in these future developments, a knowledge of biochemistry will become increasingly important. There are two main arguments to support this statement. First, the improvement of crops by agronomic or genetic means involves the manipulation of the physiological and biochemical characteristics of the plant, and it is only by an appreciation of the cellular mechanisms involved that the likelihood of success in a particular project can be guaged, the causes of failure understood, or the future possibilities identified. Secondly, the evaluation of crop varieties or changed methods of husbandry often involves the chemical analysis of material in order to assess its nutritive value, storage characteristics, or dietary acceptability.

In the title of this book the term 'biochemical analysis' is intended to describe both of these aspects of plant biochemistry and this is reflected in the arrangement of the subject matter. Chapters 2–4 deal with the technical procedures associated with the sampling and analysis of temperate and tropical crops and their products. Many of these methods are applicable to a wide range of materials including grass and other forages, cereal grains, leguminous seeds, roots, tubers, and fruits. The major constituents found in plant tissues are the subject of Chapter 5, which comments on the physiological and metabolic significance of individual compounds and also outlines the appropriate analytical techniques required for their examination. It is hoped that this dichotomous approach will enable plant biologists with only an elementary knowledge of biochemistry to initiate experiments, and at the same time develop a greater appreciation of the function and importance of the particular compounds under examination.

The historical development of biochemistry in agriculture has centred on a study of the *nutritive value* of crops and their products. There are many excellent books and reviews covering this approach and these are referred to where appropriate. Although no detailed discussion of the nutritional significance of compounds is attempted here, *selected* observations have been included, especially when they appear to be particularly topical and the subject of current investigation. Mention is also made of toxic compounds which occur in plant material. These include natural constituents and exogenous substances such as pesticides.

Many scientists have forecast changes in our current approach to food production, with a shift towards the direct exploitation of plants for human consumption, rather than for animal feeding. Undoubtedly, developments of this nature will demand a greater understanding of the mechanisms controlling the yield, composition, and quality of crop species and this is one of fields in which the biochemist will play an increasingly important part.

Although this book describes, as a continuous theme, the application of biochemical methods and principles to the scientific study of *crop* plants, it is intended that students in the pure sciences, as well as in agriculture, will find the subject matter of direct relevance to their studies, project work, or research. Each section includes a list of references which should enable the research worker to pursue a particular topic in far greater depth than is possible within this book. It is appropriate to make special mention of two standard works which are referred to extensively in the following pages. The first of these, *Modern methods of plant analysis*, edited by Paech and Tracey, was published as a series of volumes during the period 1956–64. The second, more recent work, *Phytochemical methods* (Harborne, 1973) also gives practical details of many useful techniques.

REFERENCES

HARBORNE, J.B. (1973). *Phytochemical methods.* Chapman & Hall, London.

PAECH, K. and TRACEY, M.V. (1956–64) (ed.) *Modern methods of plant analysis.* Springer-Verlag, Berin.

2. Sampling, storage, and extraction of plant material

BEFORE the initiation of a programme of biochemical analysis, great attention must be given to the proposed methods of sampling, storage, and extraction. Although it is impossible to formulate exact procedures for all possible circumstances, an awareness of the problems likely to be encountered and how to solve them is sufficient to promote a critical attitude towards these preliminary aspects of analysis.

2.1 Sampling procedures

The biochemist working with crop plants may obtain tissue for analysis in a number of different ways. The sample may be derived from a growing crop in the field, from a bulk store of grain or other harvested material, or from laboratory experiments in which the conditions of growth or storage have been carefully controlled. The approach required in each of these situations depends on the purpose of the experiment and the likely variability in composition of the material in question.

As an example, consider two separate experiments in which the amino acid lysine is to be determined. In the first experiment the analyst may wish to compare the amounts of lysine in two different varieties of a cereal, perhaps barley. To carry out this task the varieties would have to be grown at a number of different centres, in replicated plots, over a period of several years. Sub-sampling procedures would have to be carefully standardized and replicated chemical analyses would be required in order to assess error inherent in the laboratory method. Only in this way would it be possible to confirm that any small difference encountered was due to a genetic rather than an environmental determinant.

In the second experiment, the aim might be to determine the lysine content of a protein fraction isolated from plant tissue. Say the fraction represented a single molecular species, perhaps an important enzyme such as ribulose diphosphate carboxylase, present in large enough amounts to influence the nutritive value of leaf protein. In this case it can be assumed that the fraction would have a fixed composition *even when the environmental conditions of growth are varied.* Thus as long as the protein could be obtained in a purified form, the error associated with the lysine estimation would be entirely due to the *chemical* procedures employed, and replication would be necessary only in order to assess the extent of this particular source of error. This illustration shows that by applying the accepted dogma of modern biology, in this case by assuming that each specific enzyme protein has a unique primary sequence, it is possible to modify sampling procedures in appropriate circumstances and so reduce the number of analyses required. In the following discussion of field and laboratory experiments it will be assumed that similar arguments have been examined prior to the finalization of an experimental design.

2.1.1. *Field experiments*

One of the main difficulties associated with the collection of material in the field is the problem of obtaining a representative sample. For biochemical work this is especially true, due to the fact that usually only very small amounts of tissue will be used in the final assay. If samples are taken from a full-scale crop before harvest, a large number will be required from many positions in the field. Small plots are easier to examine and it is possible to carry out sequential studies by harvesting several plots at each of a number of occasions during the growing season.

The total number of samples analysed should be increased if it is found necessary to work with fresh, rather than with dried and milled material. This is because in the latter case, sub-sampling can be carried out much more easily with less chance of bias.

A further major problem arises when the compounds to be analysed are metabolically labile. For example, the analysis of free amino acids or simple sugars in leaves should ideally be carried out using extracts obtained immediately after removal of the leaves from the plant. An alternative is to freeze samples with solid carbon dioxide and transfer them to the laboratory in this condition. They should then be extracted *without thawing*. These precautions are necessary because at normal temperatures many enzymes associated with the metabolism of these compounds remain active in the leaves even after the tissue has been removed from the parent plant. Less difficulty is encountered with stable components such as lignin or other cell wall materials, where samples can be stored in polythene bags and transferred to a drying oven later in the day.

2.1.2. *Greenhouse and laboratory experiments*

With experiments carried out in controlled, or semi-controlled conditions, there is generally less difficulty in effecting rapid transfer of material to the laboratory. Usually plants in growth cabinets are raised in pots or shallow trays and each of these units would normally constitute a single sample in any subsequent statistical treatment. For this reason the sampling procedures adopted in growth cabinet studies are usually worked out before the start of the experiment, the design of which may be limited by a restriction of space.

Detailed biochemical studies often involve complex analyses on many different components and it is important to remember that fluctuations in the levels of some of these components may occur in diurnal rhythm. Thus it is sometimes important to standardize the time of day at which samples are taken.

Usually it is the aerial part of the plant which is required, and samples are easily removed by excising the shoot system just above soil level. However, root samples of growing plants may also be under examination, in which case the removal of surface contamination is often extremely difficult. Plants grown in sand, using liquid nutrients, are well suited for root studies as the sand particles can be freed from the roots by careful washing in cold water.

2.2. Storage of samples

Samples must be stored in such a way that changes in chemical composition are avoided. The method of storage should always be chosen with due regard to the nature of the eventual analyses to be carried out. Changes in composition can be due to the action of enzymes, either occurring naturally in the cells of the sampled tissue, or produced by a microbial population present in the sample. Oxidative changes may also occur due to the reaction of molecular oxygen with compounds such as the unsaturated fatty acids. A less common problem is the loss of low boiling point constituents due to volatilization; this occurs if the conditions of drying are inappropriate. Individually, each of these processes can be minimized, but occasionally a storage procedure which prevents one unwanted event may favour a second, deleterious process. In such cases

TABLE 2.1
Occurrence and prevention of chemical changes in stored samples

Process	Example	Means of reducing rate of process
Oxidative deterioration	Autoxidation of unsaturated fatty acids	Storage under nitrogen
Metabolic transformations	Breakdown of proteins by proteolytic enzymes	Storage at low temperature or rapid drying
Microbial activity	Utilization of stored sample as a substrate for the growth of bacteria or fungi	Storage at low temperature or rapid drying
Volatilization	Loss of volatile fatty acids	Avoidance of heat treatment and storage in sealed containers

separate samples may be required for particular analyses. The application of different storage procedures in the prevention of the deterioration of samples is outlined in Table 2.1.

2.2.1. *Low temperature storage*

Fresh plant material may be preserved by freezing at −12°C to −18°C using plastic self-seal bags or plastic boxes as containers. Freezing should be carried out as quickly as possible and for this reason small samples are preferred. The rate of freezing can be increased by placing tissue into liquid nitrogen or by freezing with solid carbon dioxide. Deep-freeze cabinets of a commercial type, particularly those with access from above, are suitable repositories for samples, but trial experiments should be carried out to check on the long term stability of the biochemical components which are under examination. Deep-freeze rooms are useful for large scale storage but it should be noted that often a 'de-frost' period is programmed each day for this type of installation and the temperature may rise to an unacceptable level.

Storage at low temperatures above freezing point is unsatisfactory, as many biochemical transformations continue even at 2–3°C. Some physiological processes are even accelerated under these conditions. Samples which are allowed to thaw and are then re-frozen should also be treated with caution. Freezing usually results in a loss of the permeability properties of the membranes of the cell and, upon thawing, enzymes and substrates previously in separate sub-cellular compartments may be brought into contact.

2.2.2. *Drying of samples*

For the routine examination of leafy material with respect to major constituents such as fibre, fat, nitrogen, and ash, the usual procedure is to dry the sample in a forced draught oven, at around 105°C, over a period of several hours or perhaps overnight. The sample can then be milled to a powder, the fineness of which can be checked by sieving. However, when certain plant metabolites are being studied, rapid drying may produce unwanted changes and, of course, in the case of enzyme assays such treatment would be completely inappropriate. However, if drying can be accomplished efficiently then this approach to sample preservation has wide applications and material may be stored for long periods of time. Indeed, Harborne (1973) cites reports of the analysis of essential oils of dried herbarium material dating back to the eighteenth century.

In addition to the usual form of laboratory oven equipped with a fan and an air venting arrangement, microwave ovens, which offer an alternative and extremely rapid means of drying samples are also available. However, their use is not yet widespread and they offer few advantages except in special circumstances—for example, when a large number of samples must be handled in a short period of time.

One of the most useful techniques for drying biological material involves the initial freezing of the tissue followed by sublimation of the ice at reduced pressure and low temperature. This procedure is known as *freeze drying*. It is extremely useful for drying protein precipitates and similar materials, which are obtained as light fluffy powders. Commercial equipment is available for processing fairly large quantities of wet tissues, slurries, or precipitates.

2.3. Extraction of samples

Samples for analysis may consist either of post-harvest material, often in a fairly stable physiological state, or of fresh material of high water content, such as leaf tissue. During extraction, care must be taken to minimize changes in chemical composition and to ensure that the percentage recovery is satisfactory. When extracting fresh tissue with aqueous solvents, it may be difficult to break all, or nearly all, of the cells. Other reasons for low efficiency of extraction include the use of too small a volume of solvent, failure to make a sufficient number of sequential washings, the choice of an inappropriate solvent, and the loss of components by denaturation or chemical modification.

Fresh material may be dried and milled prior to extraction or it can be extracted immediately by disrupting the tissue in the solvent using some type of manual or mechanical technique. The simplest procedure is to use a conventional *pestle and mortar*, preferably of glass, with a little acid-washed quartz sand as an abrasive. *Cone-shaped pestles* which are rotated within glass mortars of the appropriate shape are also available and these can be operated by hand or by the drive from an electric motor, usually coupled to the pestle by a flexible mounting. They are suitable only for small volumes of liquid. Another alternative is the *Waring blender*, which consists of a glass container within which a rapidly rotating knife-edged propellor breaks up the tissue in the presence of an appropriate volume of solvent (Fig. 2.1). Additional techniques which are especially suitable for dried and milled material include *mechanical shaking* with solvent for a defined period of time, and continuous extraction, involving the percolation of condensed solvent in a refluxing system, through the sample contained in a porous thimble. This latter

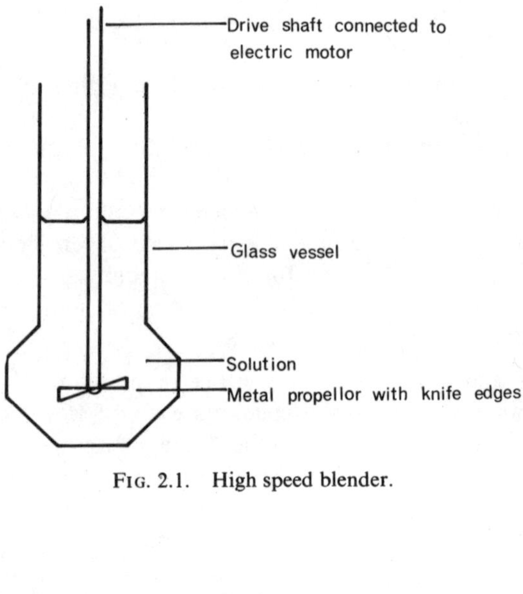

FIG. 2.1. High speed blender.

TABLE 2.2
Applicability of extraction procedures

Procedure	Organic solvents	Aqueous solvents	Disadvantages	Advantages
Mechanical shaking with solvent	Yes	Yes	Suitable only for dried and milled material	Simple and straightforward
Pestle and mortar	No	Yes	Difficult to reproduce and some cells may not be disrupted	Inexpensive apparatus and convenient to carry out
Power-driven cone pestle and mortar	Some	Yes	Suitable only for small samples	High shearing forces and good cell breakage
Waring blender	Yes	Yes	May cause frothing of solvent	Suitable for large samples
Reflux of solvent in presence of sample	Yes	No	Not suitable for fresh material	Very efficient

procedure, which is applicable only to organic solvents, usually involves the use of the *Soxhlet apparatus*. These procedures, together with comments concerning their applicability, are listed in Table 2.2.

2.3.1 *Aqueous solvents*

Occasionally, water itself is used as an extraction medium without the addition of any solutes. For example, the extraction of nitrate from dried plant material is usually accomplished by shaking a weighed sample with distilled water. However, some attempt is often made to control the pH and tonicity of the solvent, particularly when extracting enzyme proteins from fresh material. The problems associated with the extraction of enzymes are discussed in section 5.2.4. When the intention is to remove as much as possible of a particular enzyme from the tissue, a low molarity buffer likely to promote the osmotic lysis of subcellular organelles is normally employed. Tris buffer (0·04 M; pH 7·8) would be appropriate as a starting point in the development of a suitable extraction medium in these circumstances. On the other hand, if there is a requirement to maintain the structural integrity of the subcellular organelles, the osmotic strength of the solution must be increased by the addition of sucrose at a concentration of around 0·4 M. Other additives which may be employed as a means of preserving subcellular structures include certain commercial products such as 'Ficoll', which is synthesized by the co-polymerization of sucrose and epichlorohydrin. These compounds are uncharged and of high molecular weight, and give solutions of low osmotic strength at a given concentration. They act as stabilizing agents in the extraction of proteins and in the preparation of subcellular fractions (Honda, Hogladorum, and Laties 1966).

It is often advisable to carry out extractions at low temperature in order to minimize enzyme activity. Buffers, glassware, and other containers should therefore be stored in the refrigerator before use. Certain electrically powered homogenizers have an outer container and this can be packed with an ice–water mixture, so cooling the slurry or brei as it is broken—up by the rotating blades.

One disadvantage of power driven homogenizers is a tendency for the mixture under treatment to froth at its surface, with a resultant deleterious effect on proteins in solution. If denaturation occurs as a result of frothing this inevitably leads to a loss of enzyme activity.

2.3.2. *Organic solvents*

Organic solvents differ in their polarity and thus in their suitability for the extraction of different components. Harborne (1973) recommends ethanol as a good all-purpose solvent in preliminary studies of phytochemical constituents. Mixtures of ethanol and water are suitable for the extraction of many different classes of compounds. The amount of water in the mixture can be modified, according to the water content of the sample, in order to arrive at an appropriate final concentration of ethanol. Light petroleum (b.p. 40–60°C) is useful for extracting non-polar lipids but mixtures of chloroform and methanol are more suited to the removal of phospho- and glycolipids. A mixture of $CHCl_3$:MeOH (2:1) has the advantage of remaining as a single-phase system even in the presence of small amounts of water derived from fresh samples.

Further specific recommendations relating to the use of organic solvents are given in Chapter 5.

REFERENCES

HARBORNE, J.B. (1973). *Phytochemical methods*. Chapman & Hall, London.

HONDA, S. I., HOGLADORUM T., and LATIES, G. G. (1966). A new isolation medium for plant organelles. *J. exp. Bot.* **17**, 460.

3. Separation methods

QUANTITATIVE analysis in biochemistry is usually preceded by a series of fractionation steps, chosen to remove components other than those of primary interest. This chapter deals with separation techniques employed for this purpose. Some procedures have a wider application, being used both as a means of fractionation and also of quantitative assay. For example, gas–liquid chromatography (g.l.c.) can be used as a preparative tool for the isolation of small amounts of pure compounds, or it can be employed as an accurate, sensitive, and quantitative technique for measuring the amounts of individual components of complex mixtures. Thin layer chromatography (t.l.c.) can also be used as a quantitative method but only with much greater difficulty and less precision. Because of the problems associated with the quantification of t.l.c. it is treated here as a method usually employed only for the purpose of separation. Gas–liquid chromatography, on the other hand, is usually regarded as one of the most satisfactory quantitative techniques of general application and is thus dealt with in Chapter 4.

The separation of complex organic compounds can be achieved by exploiting differences in solubility, chemical reactivity, molecular size or charge, volatility, or affinity for a particular inert stationary phase. In addition to separations at the molecular level, there are many techniques which involve the separation of subcellular organelles or molecular aggregates, such as ligno-cellulosic fibres.

Certain methods have very wide application. For example, centrifugation can be used to separate solvent phases, to sediment insoluble fibrous material, or, at very high speeds, to resolve individual molecular species. Other techniques are useful only in a very limited field.

For most components of crop plants which are likely to interest the agricultural scientist, there are well-established combinations of separation and assay procedures. However, many important advances have been made by the use of novel combinations of techniques or by the development of new procedures. Many examples of this occur in the field of plant growth compounds, where it has often been necessary to establish new bioassays which have then been employed in conjunction with specific purification techniques. Whichever particular procedure is adopted for the fractionation of plant extracts, it is always important to relate the experimental method to the final assay system. An assay of high specificity can be used with less regard to the degree of purification obtained than can one of low specificity. Other reasons for choosing the purification technique with regard to the nature of the assay include the necessity of obtaining a solution in the right concentration range and of avoiding interference by specific contaminants.

3.1. Differential solubility

At some stage, almost all procedures for the isolation of particular components exploit solubility differences between the major groups of compounds. By manipulation of pH or by the use of aqueous or non-aqueous solvents it is possible to extract selectively different categories of compounds. The use of specific solvents in preliminary extraction is discussed in Chapter 5 and general observations regarding the formulation and use of solvents were made in Chapter 2.

During the further purification of extracts, solubility differences between wanted and unwanted components may be used as a means of concentrating a certain compound in either one or the other of two immiscible solvents. Once again, pH adjustment may be necessary to ensure an appropriate fractionation. An example is the extraction of free fatty acids from a total lipid extract. Free fatty acids can be separated from esterified acids in a petroleum solution by extracting with 1 % sodium carbonate. To facilitate the extraction of the free fatty acids back into an organic solvent, the carbonate solution is then acidified to suppress the ionization of the acid group. During this transfer hydrophilic compounds remain in the aqueous phase and thus the fatty acids are freed of water-soluble impurities. In this example the two purification steps involve the differential partition of solutes between two immiscible solvents. Solubility differences can also be exploited by the use of a single solvent which dissolves certain components leaving others as an insoluble residue, either on a filter paper, on the walls of a flask, or in a centrifuge tube. Finally, the selective removal of solutes from solution can be accomplished by altering the composition of the solution through changes in pH, ionic strength, or temperature. Of wide application in protein work is the selective precipitation of different protein fractions by the addition of various amounts of ammonium sulphate to the solution (Dixon and Webb 1964). This particular salt is used because of its high solubility and also because it rarely produces harmful effects on enzyme proteins. The concentration of ammonium sulphate is usually expressed in terms of percentage of saturation. By the addition of sufficient salt to give a fully saturated solution, usually all the protein present can be recovered and this is a means of obtaining the protein substantially free of water-soluble impurities.

3.2 Differential chemical reactivity

The lability of compounds in particular circumstances can be used to so modify samples that their subsequent examination with respect to specific components is facilitated. Possibly the simplest and most widely used procedure of this type is the 'ashing' of samples to remove organic matter by combustion, leaving inorganic material as a residue. Typically, a small sample is weighed into a porcelain crucible and heated in a muffle furnace at 450°C for 12 hours. The ash is then determined gravimetrically.

Another standard method involves the separation of the total lipid fraction into saponifiable and unsaponifiable portions. The saponification of plant lipids can be achieved by refluxing with a solution of potassium hydroxide which results in the release of fatty acids previously bound as constituents of complex molecules such as the phospho- and glycolipids, or the triglycerides. The unsaponifiable lipids, including the sterols, can then be separated by retention of the free fatty acids in an alkaline aqueous solution, during partition into light petroleum.

Certain cell wall components are commonly examined by the destructive removal of compounds other than those of interest. Lignin can be determined by the successive digestion of milled samples with pepsin and sulphuric acid. The organic matter in the residue is determined gravimetrically and taken as being representative of the lignin content of the sample. Other methods for the isolation of 'fibre' fractions involve a similar approach.

Despite these several important examples, the use of differential chemical reactivity is of strictly limited application. Only a few classes of compounds are sufficiently inert to survive the action of the strong reagents usually required to remove other diverse compounds.

3.3. Centrifugation

The technique of centrifugation is one of the most widely used in biochemistry. It can be used to separate solvent phases and selectively to remove insoluble particulate material.

Under the influence of an applied gravitational field, particles sediment at a rate determined by their size and density. The sedimentation coefficient S is an expression of the rate of movement of a particle in a given centrifugal field and is proportional to the difference in the densities of the particle and the medium $(\triangle\rho)$, and to the square of the radius of the particle:

$$S \propto \frac{r^2 (\triangle\rho)}{\eta},$$

where r = radius of particle and η = viscosity of medium.

Usually the sedimentation coefficient is reported as the value obtained when the temperature is 20°C and the liquid medium has the density and viscosity of water. This value, measured in *Svedberg*

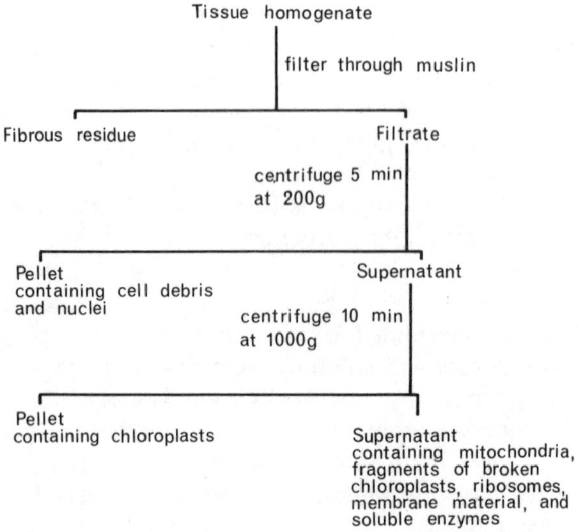

FIG. 3.1. Centrifugation of a leaf homogenate.

units, is characteristic of the particle. Thus different subcellular components can be separated according to their size; at a given speed of centrifugation, only particles above a certain size move quickly enough to reach the bottom of the tube before the experiment reaches its conclusion. The preliminary fractionation of a leaf homogenate is illustrated in Fig. 3.1. Providing that a suitable buffer is chosen for the homogenization step, this procedure is suitable for the bulk isolation of chloroplasts. Further centrifugation to sediment the mitochondria would give a rather contaminated fraction with many small fragments of broken chloroplasts also present. For most practical purposes the preparation of a mitochondrial fraction is feasible only with non-green tissue, and usually etiolated material or naturally non-photosynthetic tissue is chosen.

In crop science the practice of cell fractionation is of rather limited application although studies of the mode of action of herbicides or their analogues may involve the preparation and examination of subcellular fractions. Usually the centrifugation of samples is undertaken for the purpose of separating unwanted components prior to a particular method of quantitative analysis.

TABLE 3.1

Types of preparative centrifuge and their characteristics

	Maximum relative centrifugal acceleration (*g*)	Tube volume (ml)	Temperature control	Applications
Bench models	2000	5–100	No	Separation of solvent phases, removal of coarse precipitates, cell debris and fibrous material
Free standing, general purpose models	30 000	5–1500	Yes	Large scale sedimentation of proteins, cell debris and other fractions; recovery of plastids from cell homogenates
High speed models with evacuated rotor bowls, often referred to as 'ultracentrifuges'	150 000	5–100	Yes	Sedimentation of ultra-fine precipitates, subcellular organelles, membrane fragments, and free and bound ribosomes.

Many different types of centrifuge are available but they can be roughly classified into three groups as shown in Table 3.1. The most elaborate models operate with close control of temperature even at high rotor speeds. This is achieved by evacuating the air from the rotor bowl to reduce the generation of heat by friction. Most bench models are unsuitable for cell fractionation work as usually there is no method of maintaining solutions at low temperature during centrifugation. However, apart from this problem, they are of very wide application in general analytical work and have the advantage of low cost.

Centrifugal force is usually expressed relative to the earth's gravitational acceleration g. High speed models can provide accelerations of up to 150 000 g, or more. The force exerted on the particles in a suspension depends on the position of the centrifuge tube in relation to the axis of rotation and on the speed of revolution.

To calculate the relative centrifugal acceleration G the following equation may be used (Mahler and Cordes 1969):

$$G = 1{\cdot}11 \times 10^{-3} R \times N^2,$$

where N = revolutions per minute and R = radius of revolution in metres. The force in newtons (F) exerted by a unit mass in the centrifugal field may be calculated from $F = 9{\cdot}8\ G$.

By using different rotors and by control of the speed of the centrifuge a wide range of G values can be obtained from a single machine. The values given in Table 3.1 are approximate, and are intended only as a rough guide to the performance obtainable from different types of centrifuge.

In order to decide on the specific conditions of centrifugation in any particular circumstance it is often necessary to consider factors such as the volume of the sample, the types of rotor available, and the fate of the centrifuged sample in the subsequent analysis. Samples of large volume may involve the use of rotors and tubes which are suitable for use only at comparatively low speeds and which require long periods of centrifugation to achieve adequate sedimentation. Sometimes it is possible to choose the equipment with greater freedom, in which case the aim may be to use tubes convenient for the next step in the fractionation. This may involve the use of graduated tubes to facilitate solvent additions, pyrex tubes to allow for heating the sample, or tubes with fitted stoppers to allow for a period of storage.

3.4. Filtration

In certain situations the techniques of filtration and centrifugation represent alternative methods of achieving a particular separation. For example, the removal of a coarse precipitate from solution can be carried out either by centrifugation at low speed using a bench centrifuge or by conventional paper filtration. The choice in such cases is usually arrived at simply on grounds of convenience. As well as the standard procedures of gravity and suction filtration, using the normal range of papers, there now exist several new techniques which are of considerable application in biochemistry.

The filtration of homogenates of plant material through muslin or cheese-cloth is widely used as a means of removing fibrous material and intact tissue from homogenates prior to cell fractionation. As an intermediate between filtration at this level and filtration through paper, it is possible to pass homogenates through fine nylon net which will retain a large proportion of the cell debris but allow subcellular organelles to pass through in the filtrate. This procedure has been used in the rapid preparation of chloroplast fractions, which can be obtained within minutes of breaking up the tissues of the leaf. Nylon material of the appropriate closeness of weave can be obtained from manufacturers of synthetic fabrics.

A new development in paper filtration is the production of impregnated papers which make possible the separation of solvent phases. The properties of these papers are so controlled during manufacture that an aqueous phase is retained in the filter funnel whereas an organic phase passes through the paper. This technique offers an alternative to the separating funnel and has the advantage that it can be adopted for the simultaneous separation of many samples, without the continuous attention of the experimenter.

Another modern filtration technique involves the use of fine filter discs of closely controlled pore size which allow the selective removal of extremely small particles such as bacteria or subcellular organelles. These filters are operated under reduced pressure and usually at relatively low flow rates, using specially constructed filter holders and other equipment.

An important precaution when using filtration aids is to ensure that they introduce no contamination to the solutions under treatment. Problems of this kind can usually be overcome simply by a preliminary washing with the appropriate solvent.

3.5. Electrophoresis

Electrophoresis involves the separation of charged particles under the influence of an applied electrical field. The fact that migration under a given set of conditions occurs at a characteristic rate was first utilized for analytical purposes by Tiselius, in 1925. The movement of compounds in free solution is difficult to observe and the early methods of Tiselius suffer from the problem of diffusion, which must be minimized by the maintenance of a low temperature. Because it was usually the advancing front of each component of the migrating sample that was observed, the technique became known as *moving boundary electrophoresis*. Much of the original work on the electrophoresis of proteins was carried out by means of this method. However, an important development, the use of support media, led to a much wider application of electrophoresis in routine analysis. *Zone electrophoresis* is the term applied when the basic technique is modified in this way (Kunkel and Trautman 1959). One type of electrophoresis involves the use of *polyacrylamide gel* as the support medium. Because of its wide application this method will be described in detail and used as an illustration of the operation of a zone electrophoresis system.

3.5.1. *Polyacrylamide gel electrophoresis*

The separation of proteins or nucleic acids on gels of *polyacrylamide* can be accomplished rapidly, with high resolution, using only very small amounts of sample (Whipple 1964). The structure of the support medium gives rise to a molecular sieving effect as although

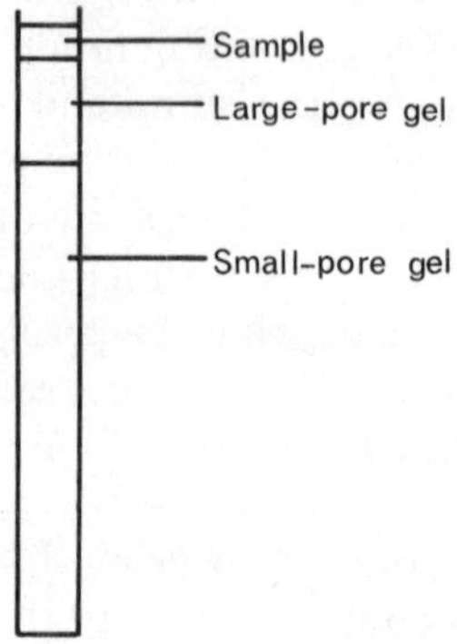

FIG. 3.2. Arrangement of gel layers for electrophoresis on cylinders of polyacrylamide.

small molecules move freely, the small effective pore size prevents the unrestricted migration of very large molecular species. The gel is formed from a monomer solution by polymerization and can either be cast in slabs or in the form of long cylinders. The latter alternative is most widely used. The monomer is mixed with an initiator solution and poured into a glass tube to form the gel. The lower and greater part of the glass tube, which is held vertically, is filled with a *small-pore gel*. A second layer, of *large-pore gel*, is then polymerized *in situ* above the first layer (Fig. 3.2). Control of pore size is achieved by altering the concentration of acrylamide.

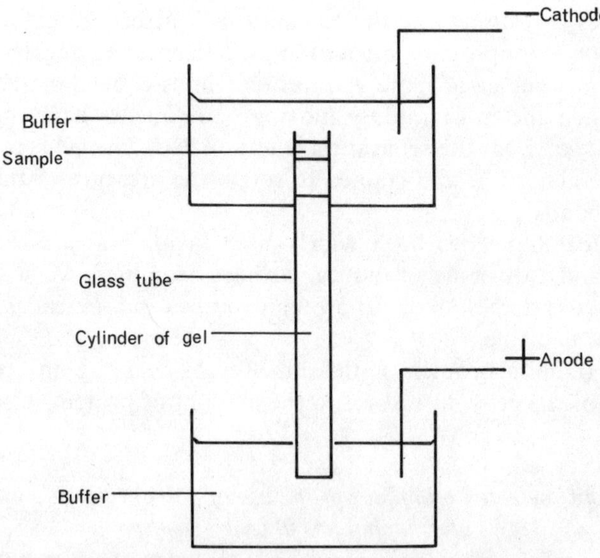

FIG. 3.3. Apparatus for polyacrylamide gel electrophoresis

To carry out the separation the gel tubes are positioned vertically in the apparatus with the lower end of the gel dipping into a reservoir of buffer, in which is located one electrode, while the other end is surrounded by buffer in a separate reservoir containing the second electrode (Fig. 3.3). When the current is switched on, charged components of a sample applied to the gel will tend to migrate towards the appropriate electrode. The rate of migration depends on the charge on each particle and to some extent on molecular size.

A number of different methods are available for applying the sample to the gel. The most straightforward is to mix the sample solution with a little sucrose to form a viscous and dense mixture. This can be applied to the top of the gel, under the surface of the reservoir buffer, using a micro-syringe. The high density of the sample prevents mixing. The current is then switched on and the components of the sample migrate rapidly into the large-pore gel. The sample tends to concentrate as it reaches the interface between the two gel layers and this is equivalent to applying the sample in a very narrow band. As migration continues into the small-pore gel the components start to separate. Proteins are usually separated at high pH and thus they tend to possess a negative charge and move towards the anode. The movement of proteins and other large molecules is restricted by the gel and this is a second factor in the separation. Lipoproteins often exist as molecular aggregates which fail to enter the small-pore gel and remain as a band at the interface. When the most rapidly moving components have migrated a distance of 5 cm the separation is terminated. The gel is removed from the glass tube and stained to reveal the presence of the component bands.

This technique has been widely used in all branches of plant science and a diversity of tissues has been examined. Most separations involve the resolution of protein mixtures and several examples are given in Table 3.2.

One common problem is the difficulty of deciding on a correct loading of the gel with respect to the amount of protein. Usually it

TABLE 3.2

Some selected examples of the use of polyacrylamide gel electrophoresis in crop science

Sample	Species	Application	Reference
Leaf proteins	Oats	Characterization of ribulose diphosphate carboxylase	Steer, Gunning, and Carr (1966)
Root proteins	Alfalfa	Biochemical evaluation of low temperature response	Gerloff, Stahmann, and Smith (1967)
Seed proteins	Wheat	Analysis of the biochemical genetics of cereal protein production	Favret, Solari, Manghers, and Avila (1969)
Fruit proteins	Apple	Study of fruit ripening and varietal differences	Clements (1970)

is necessary to run a range of samples differing in either volume or concentration. An alternative is to assay the sample solution for total protein prior to electrophoresis. The amount of protein required varies according to the number of protein components present and their relative amounts. Overloading results in a slow rate of migration, with poor resolution and heavy staining.

In order to obtain a semi-quantitative estimate of individual proteins the gels can be examined with a densitometer, an instrument which records the light absorption at each point in the gel. If the resolution of bands is good and the right wavelength is chosen, the continuous trace produced by the densitometer can be used to obtain an estimate of the amount of protein represented by each peak.

Separations of nucleic acids can also be undertaken by polyacrylamide gel electrophoresis. The gels differ from those used for protein separations with respect to their effective pore size. The application of nucleic acid separations to studies of crop plants has so far been rather limited.

3.5.2. *Electrophoresis on paper, cellulose acetate, and other media*

A very wide range of materials has been used for the purpose of zone electrophoresis. Some media have been little used after the first reports of their properties appeared in the literature. Others have been widely adopted by many laboratories and special commercial preparations have often resulted from this popularity. Support materials include paper, cellulose acetate strips, agar gels, starch gel, granular starch, polyvinyl chloride resins, and sponge rubber (Bier 1959).

Electrophoresis on paper strips soaked in buffer solution can be

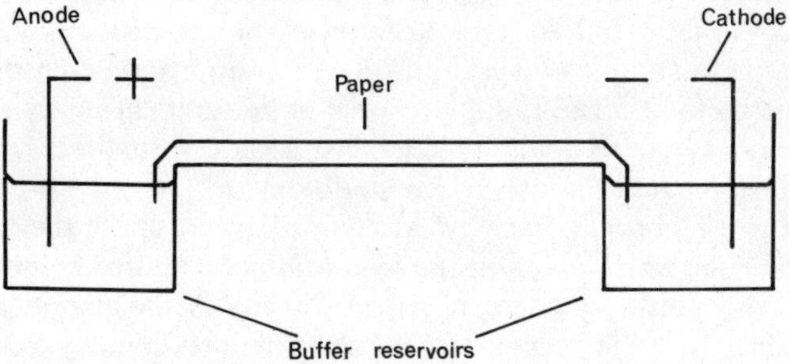

FIG. 3.4. Apparatus for electrophoresis on paper or other similar support media.

carried out in commercially available units consisting of two plastic reservoirs containing the electrodes, separated by a flat bridge to support the paper (Fig. 3.4). Good separations of alkaloids, amino acids, and other compounds can be obtained in this way.

To achieve rapid fractionation it may be necessary to increase the voltage. The separation of amino acids by high voltage electrophoresis is an excellent means of obtaining quick results. At low electrical field intensities, the evaporation of buffer from the surface of the paper produces sufficient cooling, but at high voltage special arrangements are necessary to ensure adequate temperature control. The paper may be cooled by contact with an imiscible solvent such as carbon tetrachloride, or with a metal surface, which in turn is cooled by a flow of water.

Cellulose acetate is another support material which has achieved wide usage. Apparatus suitable for paper electrophoresis can often be employed for separations on this medium. Generally the resolution of proteins is less satisfactory than that achieved with polyacrylamide gels. However, the technique is slightly more convenient as the cellulose acetate strips are purchased ready for use. Many other support materials are available but they are of limited or highly specialized application.

3.6. Chromatography

Although chromatographic methods are used in chemistry as well as in biochemistry, it is within the latter discipline that chromatography has had its greatest impact. Most types of chromatography involve mild conditions of separation and are thus ideally suited to the examination of complex biological molecules exhibiting some degree of lability. Indeed some aspects of biochemistry have been amenable to study only since the advent of appropriate chromatographic methods. The rapid advances in lipid biochemistry and in amino-acid analysis achieved in the last decade are a direct result of the development of improved methodology.

The word ‘chromatography’ was coined as a result of the earliest experiments, which involved the separation of coloured compounds whose movement could be observed directly in the analytical system. Since then, the technique has been extended to many classes of compounds, most of them colourless and thus observable only by some appropriate chemical, or other, test, after the completion of the separation.

All chromatographic methods depend on the distribution of compounds between two dissimilar phases. When the two phases are both liquids, compounds tend to concentrate in one or other of the phases according to their *partition coefficients*. If one liquid is moving relative to the other, then a compound which tends to concentrate in the *stationary phase* will remain close to the point of application of the sample, whereas a compound which tends to concentrate in the *mobile phase* will be swept away from the origin. Paper chromatography depends on the partition of compounds between a stationary water phase, consisting of water bound to the cellulose fibres, and a moving solvent phase. The sample is applied at one end of a paper strip and solvent is allowed to move over the sample spot and migrate by capillarity towards the far end. Hydrophobic compounds are swept along by the solvent while hydrophilic compounds remain nearer the starting point. This type of separation is known as *partition chromatography*.

Three other types of chromatography are widely used. The first of these, *ion-exchange chromatography* involves the use of an ionized stationary phase which selectively binds charged molecules present in the sample. Usually the stationary phase is in the form of a particulate resin which can be used to pack a glass column. The sample is applied to the top of the column and a solvent, or a series of solvents, is used to wash through the various components of the sample. This process of 'washing' the stationary phase is known as *elution*. The liquid collected at the bottom of the column is the *eluate*. Modified celluloses are also used as ion-exchangers. An example is DEAE (diethylaminoethyl) cellulose. This stationary phase can be used with organic solvents (for example, see section 5.4.3.). In ion-exchange chromatography the ions in the sample solution are held by the charged groups on the particles of the stationary phase. As solvent passes over the solid phase, weakly held ions are swept away more quickly than those which are tightly bound. In some cases complete elution of components is best achieved by the successive use of different solvents, or *eluants*. This is known as *stepwise elution*. Separations based on the ion-exchange principle can also be carried out using analytical systems other than the conventional column arrangement.

A second type of separation, *adsorption chromatography*, involves the preferential adsorption of compounds by the stationary phase, which is usually a finely divided inert material such as alumina, silicic acid, or magnesium silicate. The basis of the separation is similar to that of partition chromatography, in the sense that the '*distribution ratio*' of compounds between solid and mobile phase determines their mobility. With some techniques both partition and adsorption effects contribute to the separation.

Molecular exclusion chromatography is a modern technique depending on the availability of media which will retard compounds in a column, or some other system, according to molecular size. The most widely used of these media are the cross-linked dextrans (such as 'Sephadex') which swell in water to form a gel. Small molecules tend to be retained in the pores of the gel particles whereas large molecules are excluded from the gel matrix and are swept along by the mobile phase. *Hydroxyalkylated dextrans* are used when organic solvents are required to maintain compounds in solution. An alternative, but misleading name for this type of separation is 'molecular sieve' chromatography. Unlike a sieve, the gel retains the *small* molecules rather than the large ones.

Different grades of dextran gels are available and they are chosen according to the molecular weights of the compounds under examination. The technique is particularly suitable for separating proteins according to molecular size. In fact, if a range of standard proteins is first examined, it is then possible to obtain an estimate of the molecular weight of a sample protein. Cross-linked dextran gels are usually used as column packing materials but they can also be coated on thin-layer plates.

3.6.1. *Column methods*

In column chromatography a sample is applied as a solution to the surface of the packing material in a vertical column. Elution of components is obtained by maintaining a flow of solvent through the stationary phase. The eluant is usually collected in small measured aliquots with an automatic *fraction collector*. The successive positioning of tubes under the column outlet is triggered by the electronic sensing of drops of liquid entering each tube, by the automatic measurement of eluant volume, or simply by a timing device. In the latter case it is essential that a constant flow is maintained in order to ensure that each fraction contains the same amount of liquid. For a complex separation more than a hundred fractions may be collected. Each is then assayed by an appropriate method and the concentration in each tube is plotted against fraction number. This gives the *elution profile* of the separation. The contents of groups of tubes representing each peak in this profile are then pooled and used in further analysis. When the separation is less complex, for example when only two or three components are present, a slightly different approach can be adopted. Often stepwise elution can be used to remove first one component and then the next. This means that each fraction corresponds to a particular step in the elution procedure and will contain a single component. Usually such fractions are of large volume and the use of a fraction collector is unnecessary.

The design of the column, its proportions, and the nature of the outlet tubes and connections are of great importance in all but the simplest separations. Columns may be constructed of glass or plastic and can be surrounded by a water jacket to maintain a constant temperature. Usually the packing material is supported on a sintered glass or metal disc, or on a fine nylon net placed at the bottom of the column. The outlet connection should be of small aperture and the design of the end of the column should promote unrestricted flow, with no 'dead-volume', as this promotes broadening of the peaks in the elution profile. Commercially available columns are suitable for most purposes and the convenience of standard fittings and connecting tubes encourages their use, in preference to laboratory constructed apparatus.

The column packing can be any of a wide range of materials, depending on the nature of the separation. Some widely used examples are listed in Table 3.3.

TABLE 3.3
Some common packing materials for column chromatography

Adsorption chromatography	*Ion-exchange chromatography*
Alumina	Cross-linked phenolic, methacrylic and polystyrene resins
Silicic acid	Modified celluloses
Magnesia	Modified cross-linked dextrans
Talc	
Inuiin	
Partition chromatography	*Molecular exclusion chromatography*
Cellulose	Cross-linked dextrans
Silica gel	Hydroxyalkylated cross-linked dextrans

3.6.2. *Paper chromatography*

This technique is based on the exploitation of differences in the partition coefficients of compounds between a moving solvent phase and a stationary water phase. The water is bound to the cellulose fibres of a paper strip or sheet, which is allowed to dip into a trough of solvent. The flow of solvent may be in a descending or an ascending direction.

Samples are applied with a micro-syringe to the dry paper. The size of the sample spot is kept as small as possible. After the solvent in the sample has evaporated, the paper is placed into a chromatography tank, usually of glass, with one edge dipping into a shallow trough (Fig. 3.5). The tank is often lined with filter paper soaked in

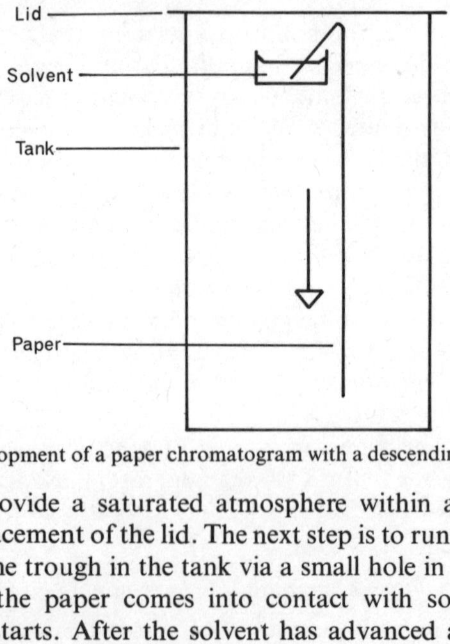

FIG. 3.5. Development of a paper chromatogram with a descending flow of solvent.

solvent to provide a saturated atmosphere within a short period after the replacement of the lid. The next step is to run the *developing solvent* into the trough in the tank via a small hole in the lid. As the trough fills, the paper comes into contact with solvent and the *development* starts. After the solvent has advanced a certain, predetermined distance, the paper is removed from the tank and dried. The separated components of the original mixture are located with an appropriate reagent sprayed onto the paper. Each compound has a characteristic rate of movement under a given set of conditions. This is expressed as an R_f value:

$$R_f = \frac{\text{distance moved by compound}}{\text{distance of solvent front from the starting point}}.$$

Thus compounds of high mobility have R_f values which approach 1·0 while for those of low mobility R_f tends to zero. The value for any particular compound can vary by as much as 10 per cent due to differences in temperature, the concentration of the sample, or other operating variables (Cramer 1954). For this reason it is usual to run reference compounds with each sample rather than rely on previously determined R_f values as a means of identification.

The scope of paper chromatography is considerable. The requirements for the successful separation and characterization of a compound have been summarized by Cramer (1954), as follows:

(a) the compound must be slightly soluble in water and in the mobile organic phase;
(b) it must be detectable on the developed chromatogram;
(c) it must not react chemically with either the paper or the solvent.

Components of plant tissues which can be separated include amino acids, sugars, organic acids, alkaloids, and the flavanoids, including the anthocyanin pigments.

Complex mixtures are sometimes incompletely resolved by development in a single solvent. In such cases it may be possible to improve the separation by carrying out a second development, using a different solvent migrating at right angles to the first. This is *two-dimensional chromatography.* Before treatment with the second solvent the square sheet of paper must be thoroughly dried. It is then positioned in a new chromatography tank and the next solvent is run into the trough to start the second development. The two solvents should be chosen with care, to ensure that compounds running together in the first solvent separate in the second. In addition, the first solvent must be sufficiently volatile to evaporate readily from the paper prior to the next stage in the separation. Two-dimensional chromatograms reveal compounds spread out over a square surface. Their position is characterized by measuring the R_f value for each solvent. These values correspond to the coordinates of a geographical map and like these coordinates they are used to describe the position of particular features.

3.6.3. *Thin layer procedures*

Thin layer chromatography (t.l.c.) involves the separation of compounds on glass or plastic plates coated with a thin layer of support material. This stationary phase may operate on either the partition, adsorption, ion-exchange, or molecular exclusion basis. The majority of separations in t.l.c. are carried out on silica gel, where the effect is mainly due to partition, as in the case of paper chromatography. The distance of migration of solvent required to give good separation is usually much less than that required for an equivalent separation on paper. The method is thus rapid and of high resolving power.

To coat the glass plates with the stationary phase a slurry is first prepared with water. In the case of silica gel, two parts of water are mixed with one part of dry powder. After mixing, the slurry is poured into a spreading device which has previously been positioned on the first of the glass plates, which are arranged end to end on a horizontal surface. As the spreader is moved over the surface of the plates a thin, uniform film of slurry is left behind. Care must be taken to move the spreader at a steady speed in order to avoid irregularities in the gel surface. After drying for a short period at room temperature, the plates can be activated by heating in an oven at 110°C, usually for around 30 minutes.

If the special equipment used in the thin layer technique is not available, pre-coated plastic sheets can be purchased, or glass plates can be coated in an alternative manner. One procedure suitable for small plates, such as glass microscope slides, is simply to dip the glass into a slurry, withdraw it and allow to dry in a horizontal position. The slurry on one side and along the edges is then removed, and the thin layer is ready for use. Layers formed in this way tend to be rather irregular and they are suitable only for the least critical tasks, such as monitoring the eluant from a packed column. Another method involves the use of plastic adhesive tape which is placed around the outside of one surface of the plate. A little slurry is then poured into the centre of the plate and is spread by rolling a glass rod backwards and forwards along the plate with its ends resting on the raised surface of the tape. After drying the tape is removed and the plate is ready for use. The thickness of the layer can be controlled by the choice of tape. In general, these methods are less satisfactory than the use of a well-designed spreader of commercial design.

The sample is applied either as a spot, using a micro-syringe, or as a streak, using an automatic sample applicator. In each case, great care must be taken to avoid damaging the surface of the stationary phase. A useful technique is to apply the sample with a micro-syringe fitted with a long, flexible needle of the type used for gas–

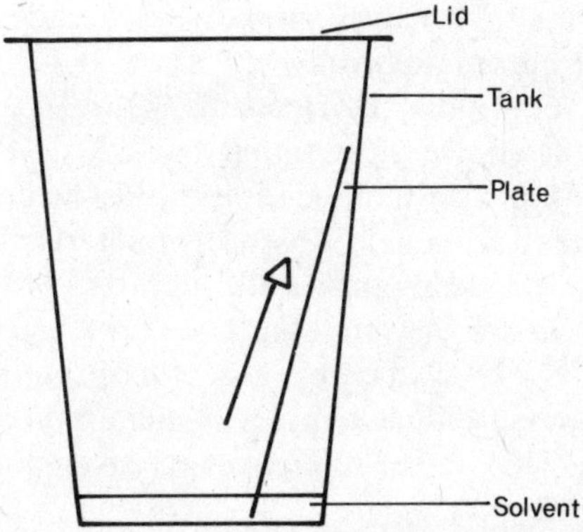

FIG. 3.6. Development of a thin-layer chromatogram.

liquid chromatography. The needle can be carefully held in position with one hand while the other is used to operate the plunger.

Thin layer plates are usually developed by standing the glass or plastic sheet vertically in a glass chromatography tank containing solvent to a depth of ~1 cm (Fig. 3.6). The solvent migrates up the surface of the thin layer due to capillarity, and components of the sample are either swept along by the solvent, or retained to a greater or lesser degree by the stationary phase. Usually, a reasonable separation is achieved after the solvent has moved 10 cm. Pre-coated plastic sheets can be cut with scissors into strips, and these can be developed using a conical flask fitted with a foil cap, in place of a chromatography tank. Excellent separations can be achieved with this simple arrangement.

After development the plates are dried and the components of the sample are located by spraying with an appropriate reagent. If the reagent does not react with the compounds under examination, individual spots can be scraped from the plate using a spatula to dislodge the surface coating. After removing the compound in question from the support material by shaking with solvent, further analysis can be carried out. If the only available spray reagents cause chemical breakdown of the sample, then part of the plate can be sprayed while the rest is shielded. The protected portions of the separation can then be used for further study, the location of components being assumed from an examination of the treated areas.

Complex mixtures may not separate completely in a single solvent and two-dimensional separations may be required. This adaptation of the basic technique involves similar considerations to those discussed for paper chromatography.

In order to achieve satisfactory results with t.l.c., it is essential to combine the right choice of stationary phase, developing solvent, and spray reagent. In addition the sample must be in a suitable form as regards concentration and absence of interfering compounds. There are many detailed reports in the literature of methods applicable to various classes of compounds. As a primary reference. the laboratory handbook on t.l.c. by Stahl (1965) should be consulted. In Table 3.4 some specific examples of the application of t.l.c. in crop science are listed. Other references are given in the appropriate parts of Chapter 5.

TABLE 3.4

Some selected applications of thin layer chromatography in crop science

Materials under examination	Application	Reference
Alkaloids in plant extracts	Analysis of toxic principles	Bull, Culvenor, and Dick (1968)
Chlorophyll degradation products	Evaluation of post-processing changes in green vegetable crops	Buckle and Edwards (1969)
Lipids labelled with carbon-14	Study of fruit ripening	Galliard, Rhodes, Wooltorton, and Hulme (1968)
Leaf surface waxes	Investigation of leaf surface chemistry	Hamilton and Power (1969)
Flavour compounds in hops	Analysis of bitter principles in hops for brewing	Stahl and Schorn (1965)
Seed oils	Evaluation of alternative oil producing plants	Sengupta, Roychoudhury, and Saha (1974)
Plant hormones	Identification of gibberellins	Cavell, Macmillan, Pruce, and Sheppard (1967)
Insecticides	Metabolism of exogenous compounds by plant tissues	Yu, Kiigemagi, and Terriere (1971)

3.6.4. *Preparative gas chromatography*

Gas or gas–liquid chromatography (g.l.c.) can be used as a preparative tool for the isolation of small quantities of pure compounds, which can then be used for further study. The main disadvantage with the technique is the restriction on sample size imposed by the design of the equipment. The essential features of gas chromatography are discussed in section 4.2.1. Briefly, the apparatus involves a column operated at an elevated temperature to maintain components in the gas phase, coupled at its output end to a detection system. In preparative work a *stream-splitting device* is incorporated into the apparatus, at the end of the column. Part of the gas flow from the column is vented to the atmosphere and a smaller part is directed into the detector. When a particular component is detected the vented gas stream is diverted into a trapping device, usually consisting simply of a glass tube maintained at low temperature. This causes the condensation of the compound which can then be eluted from the tube with solvent. A succession of traps can be used to collect each component of a sample. The compounds which are collected can be used for infrared spectroscopy or for examination by other techniques.

REFERENCES

Books and Review articles

BIER, M. (1959). *Electrophoresis*. Academic Press, New York.

BULL, L. B., CULVENOR, C. C. J., and DICK, A. T. (1968). *The pyrrolizidine alkaloids*, North Holland Publishing Company, Amsterdam.

CLEMENTS, R. L. (1970). Protein patterns in fruits. In: *The biochemistry of fruits and their products* (ed. A. C. Hulme). Vol. 1, p. 159. Academic Press. New York and London.

CRAMER, F. C. (1954). *Paper chromatography*. Macmillan, London.

DIXON, M. and WEBB, E. C. (1964). *Enzymes*. Longmans, New York.

KUNKEL, H. G. and TRAUTMAN, R. (1959). Zone electrophoresis in various types of supporting media. In: *Electrophoresis* (ed. M. Bier), p. 226. Academic Press, New York.

MAHLER, H. R. and CORDES, E. H. (1969). *Biological chemistry*. Harper and Row, New York, Evanston, and London.

ORNSTEIN, L. (1964). Disc electrophoresis—I. Background and theory. *Ann. N.Y. Acad. Sci.* **121**, 321.

STAHL, E. (1965). *Thin layer chromatography*. Springer-Verlag, Berlin.

STAHL, E. and SCHORN, P. J. (1965). Hydrophilic constituents of plants. In: *Thin layer chromatography* (ed. E. Stahl), p. 371. Springer-Verlag, Berlin.

STEER, M. W., GUNNING, B. E. S., and CARR, D. J. (1966). Fraction I protein from oat leaves. In: *Biochemistry of chloroplasts* (ed. T. W. Goodwin), Vol. 1, p. 285. Academic Press, London and New York.

WHIPPLE, H. E. (1964). Gel electrophoresis. *Ann. N.Y. Acad. Sci.* **121**, Article 2 (collected reprints), 305–650. (Review article)

UNIV. COLL. N.W. BANGOR LIBRARY

Research papers

BUCKLE, K. A. and EDWARDS, R. A. (1969). Chlorophyll degradation products from processed pea puree. *Phytochemistry* 8, 1901.

CAVELL, B. D., MACMILLAN, J., PRUCE, R. J., and SHEPPARD, A. C. (1967). Thin layer and gas–liquid chromatography of the gibberellins; direct identification of the gibberellins in a crude plant extract by gas–liquid chromatography. *Phytochemistry* **6**, 867.

FAVRET, E. A., SOLARI, R. MANGHERS, A. and AVILA, A. (1969). Genetic control of the qualitative and quantitative production of endosperm proteins in wheat and barley. *Proc. FAO/IAEA*. New approaches to breeding for improved plant protein, 87.

GALLIARD, T., RHODES, M. J. C., WOOLTORTON, L. S. C. and HULME, A. C. (1968). Metabolic changes in excised fruit tissue. *Phytochemistry* **7**, 1453.

GERLOFF, E. D., STAHMANN, M. A., and SMITH, D. (1967). Soluble proteins in alfalfa roots as related to cold hardiness. *Pl. Physiol., Lancaster* **42**, 895.

HAMILTON, R. J. and POWER, D. M. (1969). The chemical composition of the surface wax of *Lolium perenne*. *Phytochemistry* **8**, 1771.

SENGUPTA, A., ROYCHOUDHURY, S. K., SAHA, S. (1974). The triglyceride composition of *Hibiscus esculentus* seed oil. *J. Sci. Fd Agric.* **25**, 401.

YU, S. J., KIIGEMAGI, U., and TERRIERE, L. C. (1971). Oxidative metabolism of aldrin and isodrin by bean root fractions. *J. agric. Fd Chem.* **19**, 5.

4. Assay procedures

4.1. Optical methods of analysis

Many important techniques in biochemistry involve the measurement of light intensity. In order to relate an optical measurement to the amount of a particular component in the sample, it is necessary to exploit the ability of atoms or molecules to absorb or emit light when exposed to a particular set of conditions. An important requirement is that the effect observed is characteristic of the subject of the assay. Optical methods of analysis involve the use of special equipment, which may be simple and inexpensive, or extremely complex, with design features involving the latest technological developments.

The most straightforward procedure, *colorimetry*, involves the measurement of the absorption of light by a coloured solution. A development of this technique depends on the use of a *spectrophotometer*. In each of these methods it is the absorption of light by

TABLE 4.1
Optical methods of analysis

Method	Basis of the technique	Apparatus
Molecular absorption analysis		
(a) Colorimetry	Absorption of visible light by molecules in solution	Colorimeter
(b) Spectrophotometry	Absorption of light of a *narrow wavelength band* by molecules in solution	Spectrophotometer
Atomic absorption analysis	Absorption of light by a free atomic species	Atomic absorption spectrophotometer
Atomic emission analysis	Emission of light by atoms in an excited electronic state	(a) Flame photometer (b) Atomic emission spectrophotometer
Fluorimetry	Fluorescence of a solution	(a) Fluorimeter (b) Spectrofluorimeter

molecules which is measured. The fluorescence of compounds in solution is measured with a *fluorimeter* or a *spectrofluorimeter*. Free *atomic* species can be estimated by measuring the emission or absorption of light under carefully controlled conditions. The main characteristics of these methods are listed in Table 4.1 and they are described in greater detail in the following pages.

4.1.1. *Molecular absorption analysis*

The absorption of light by a compound in solution increases with the concentration of the compound, and this effect is exploited in colorimetry and in spectrophotometry. Coloured compounds, such as the chlorophylls, can be estimated directly while other compounds which give a colour reaction with a particular chemical reagent, can also be assayed in this way.

The absorption of light by solutions is described by a combination of the laws of Beer and Lambert which relate absorption to concentration and to the thickness of the absorbing layer respectively. Algebraically, these statements are summarized as follows:

$$\log \frac{I_0}{I} = \mathrm{Ecl},$$

where E = molar extinction coefficient,
c = concentration in moles per litre,
l = distance that light travels through the liquid,
I_0 = intensity of the incident light,
I = intensity of the emergent light.

The expression $\log I_0/I$ is referred to as the *absorbance* or *optical density*.

The *colorimeter* consists of four major components. A *light source* is located immediately behind a coloured *filter* which allows light of a restricted, but fairly wide, wavelength band, to impinge on a glass *cuvette* containing the sample solution. Light passing through the solution is detected by a *photoelectric cell* (Fig. 4.1). In order to measure the optical density of a solution the meter reading is first adjusted to 100 per cent transmission (zero absorption) with a 'blank' in position (in the case of an aqueous solution, the blank would consist of the cuvette containing distilled water). The sample is then inserted in place of the blank and the optical density read directly from the calibrated meter, which records the signal from the

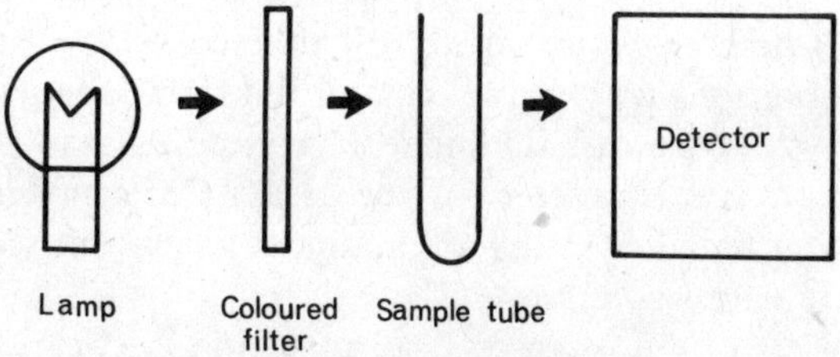

FIG. 4.1. Colorimeter.

photoelectric cell. The concentration of the solute can then be calculated using the equation given above. If the coloured compound is a product of the chemical reaction of the sample and a reagent, it is usual to avoid the use of the molar extinction coefficient, which may not be known, by plotting a calibration graph using standard solutions. The concentration corresponding to the optical density of the sample is then obtained from this graph (Fig. 4.2).

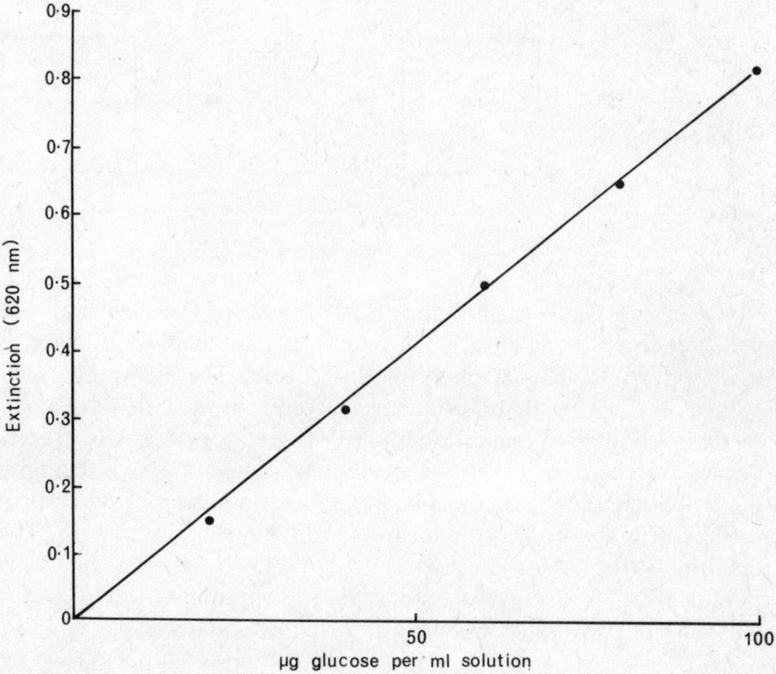

FIG. 4.2. An example of a calibration graph for use in spectrophotometry; in this case a plot obtained with the anthrone method for total sugars (section 5.5). The line related extinction at 620 nm to the concentration of glucose.

The colorimeter is an ideal instrument for routine analyses involving solutions with a single major light-absorbing component. The filter system is adequate under these conditions to provide light of an appropriate character for the assay. Colorimeters are often used in automated systems of analysis where the conventional cuvette is replaced by a *flow-cell*.

In order to obtain greater discrimination between compounds with overlapping absorption spectra the *spectrophotometer* is used. The filter system which is used in the colorimeter to obtain light of a particular colour is replaced by a prism or diffraction grating, together with an adjustable aperture regulating the width of the wavelength band. The region of the spectrum utilized for a particular assay is selected by rotation of the prism or grating. The light passing through the sample solution is detected by a photomultiplier tube (Fig. 4.3). With most instruments measurements can be made

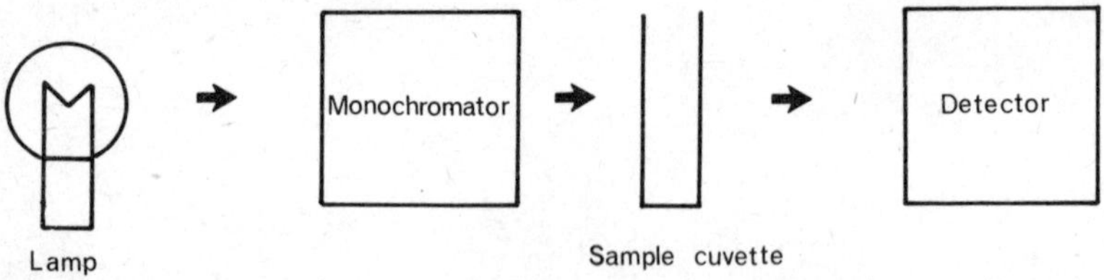

FIG. 4.3. Spectrophotometer.

in the ultraviolet as well as the visible region of the spectrum, but for these lower wavelengths silica cuvettes must be used in place of glass, which absorbs strongly in this region. The advantages displayed by the spectrophotometer in comparison with the colorimeter are improved resolution and sensitivity, together with greater versitility. The more elaborate spectrophotometers have a scanning device which allows for the automatic determination of the optical density of a solution as a continuous function of wavelength. The resulting *absorption spectrum* is traced out by a recorder. This modification is useful for identifying compounds and also for choosing the optimum wavelength for a particular assay.

4.1.2. *Atomic absorption and emission analysis*

In atomic emission analysis the presence of atoms in an excited electronic state leads to the emission of electromagnetic radiation.

This radiation occurs at a characteristic wavelength and its measurement offers a means of estimating a range of metallic elements. Usually excitation is achieved by passing the sample, as a fine spray, into a flame. With solutions containing a single compound the colour of the flame is easily assessed by the naked eye, a procedure exploited in qualitative inorganic analysis. For the quantitative examination of sample extracts, a light-filtering system is used to select a particular wavelength of light. A photodetector then senses the light and produces a signal which can be related to the concentration of the sample (Fig. 4.4). The element most commonly estimated by an atomic emission procedure is potassium.

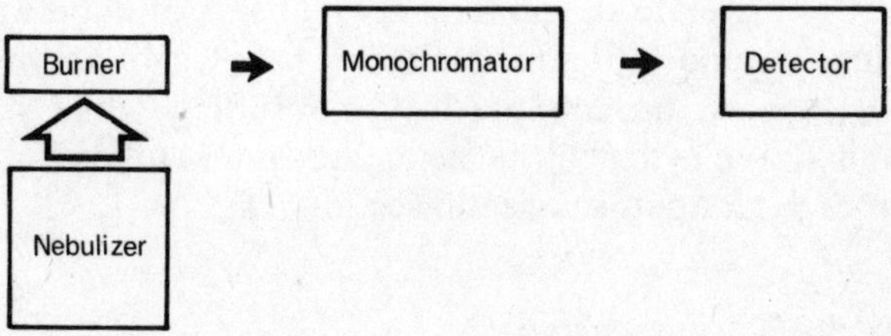

FIG. 4.4. Atomic emission spectrophotometer.

In atomic absorption analysis, the *absorption of light* by free atomic species is measured. A flame system is generally employed to dissociate elements from their chemical bonds. The atoms absorb light at a characteristic wavelength when present in their unexcited or *ground state*. A mixture of air and acetylene produces a flame which is of a sufficiently high temperature to ensure the presence of free atoms of most elements. The use of nitrous oxide in place of air results in a higher temperature and this is necessary for the estimation of certain elements. The sample solution is passed into the flame via a *nebulizer*, which forms a fine spray or mist. The source of light is a hollow cathode lamp containing the element under examination. The lamp produces radiation at the appropriate wavelength for absorption by the free atoms of the sample. After passing through the flame the light impinges on a *monochromator* which removes unwanted emissions but transmits light of the test wavelength which passes to a *photodetector* (Fig. 4.5). The latter device produces a signal which allows for the measurement of light intensity and hence of absorption caused by the sample. Elements which are determined by atomic absorption analysis include cal-

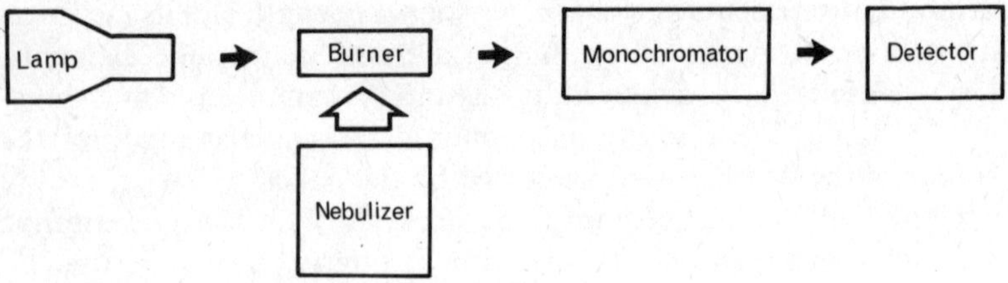

FIG. 4.5. Atomic absorption spectrophotometer.

cium, cobalt, copper, lead, magnesium, manganese, molybdenum, potassium, selenium, sodium, and zinc.

In order to estimate the mineral components of plant material it is necessary to remove organic matter by heat treatment. Either *wet* or *dry ashing* can be employed. The resulting white powder is dissolved in dilute hydrochloric acid, filtered and taken for analysis. These procedures are discussed in section 5.8.

4.1.3. *Spectrofluorimetry*

When solutions of certain compounds are exposed to a light source they absorb energy and subsequently re-emit a part of this energy as fluorescence. Many organic compounds exhibiting this characteristic absorb light in the ultraviolet region of the spectrum and emit fluorescence at a wavelength corresponding to part of the visible spectrum. The measurement of the intensity of the fluorescence can be used as a means of quantitative assay. In all the optical methods discussed previously in this section, the light source, sample, monochromator, and detector, can be regarded as forming a functionally linear succession of modules. In the spectrofluorimeter the detector is usually positioned to receive light at right angles or at some other, smaller angle to the incident beam. A monochromator is used to select a particular wavelength band and in this way interference by compounds other than the one under examination can be reduced. Filters may also be required to reduce effects due to scattered ultraviolet light from the source. The intensity of the fluorescence is influenced by temperature, the nature of the solvent used and sometimes by its pH, and also by the presence of contaminants which quench the emitted radiation.

One of the most important applications of spectrofluorimetry is the estimation of certain B-group vitamins. However, many other compounds including chlorophyll, coumarin, and certain alkaloids can be examined in this way.

4.2. Quantitative chromatography

Chromatographic procedures which have been adopted as routine quantitative methods usually involve a detection module positioned at the output side of the separation system.

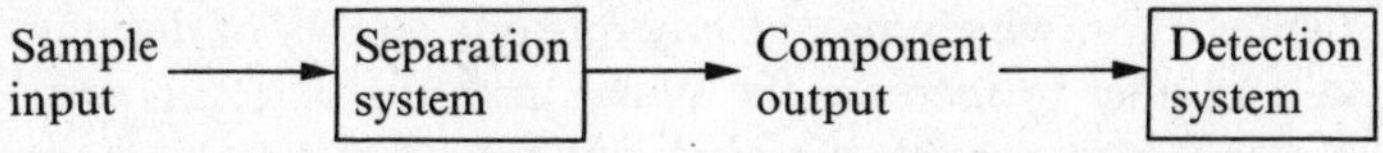

In practice, it is impossible to achieve this arrangement with techniques such as paper or thin layer chromatography, and for this reason these methods are used mainly for qualitative purposes. In contrast, gas phase chromatography offers the possibility of detecting components of a mixture as they pass some fixed point in the apparatus. Liquid chromatography offers the same advantage to the analyst.

4.2.1. *Gas and gas–liquid chromatography*

Both these methods involve a moving gas phase which tends to sweep components of the sample towards the column outlet, and a stationary phase, which is either a liquid retained on an inert support material (gas–liquid chromatography, g.l.c.), or a solid which itself constitutes the bulk of the column packing (gas chromatography). Separations are usually carried out at temperatures between 80 °C and 300 °C, but lower temperatures are used for compounds which

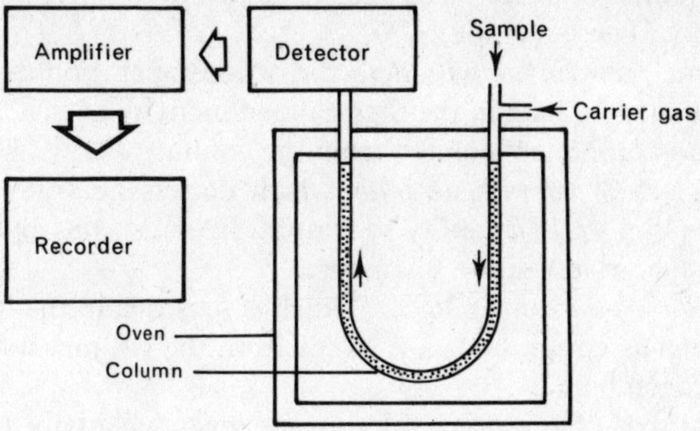

FIG. 4.6. Gas chromatograph.

are gaseous at room temperature, including the important plant hormone, ethylene. Samples are introduced through a self-sealing rubber septum using a syringe. Sample volumes range from 1 μl to 100 μl, except in the case of gaseous samples which may be as great as 15 ml. The arrangement of a typical chromatograph is shown schematically in Fig. 4.6. The apparatus consists of four functional modules:

(a) the packed column maintained in a thermostated oven;
(b) the detector, which may be any one of a variety of designs;
(c) an amplifier to increase the signal from the detector; and
(d) a recorder to provide a permanent record of each component.

The flow rate of the carrier gas, which is usually either nitrogen or argon, is carefully regulated. This means that as long as the oven temperature remains constant the time taken to elute a particular component, the *retention time*, is characteristic of that particular compound.

Columns are usually made of glass although stainless steel is required at high operating temperatures. Often the column is a metre or more in length and is accommodated in the oven in the form of a coil. The bore of the column is usually 0·4 cm. In g.l.c. a packing material such as diatomaceous earth is coated with the liquid phase at a fixed percentage of its weight. Prepared stationary phase materials and packed columns are both available from commercial sources. For certain special purposes, *capillary* or '*open-bore*' columns are used. These are of very small diameter and the liquid phase is simply coated on the internal surface of the capillary, which may be up to 100 m in length. No packing material is used.

Many different detector designs have achieved fairly widespread acceptance. These include:

(a) the *thermal conductivity detector*, whose operation depends on differences arising in the thermal conductivity of the carrier gas due to components eluted from the column;
(b) the *electron capture detector* which detects neutral molecules according to their ability to capture free electrons as they pass through an ionization chamber;
(c) the *flame ionization detector* which is sensitive to the formation of ions as compounds are swept from the column into an air/hydrogen flame.

For a number of reasons, including its great sensitivity to carbon compounds, the flame ionization system is the most widely used of these designs in biological studies. The electron capture detector is extremely useful for certain special purposes, including the estimation of halogen-containing pesticide residues in plant materials.

When selecting a gas chromatograph for a particular purpose, it is essential to choose the correct combination of detector, type of column, solid or liquid stationary phase, and operating temperature. Fortunately, most modern chromatographs consist of a series of individual modules and this affords great flexibility. Details of appropriate combinations for different separations are given in Chapter 5.

4.2.2. *Liquid chromatography*

The easiest method of obtaining quantitative results is simply to collect discrete fractions and examine these individually by some procedure such as gravimetric analysis. However, automated systems are now available and these have been used to examine a wide range of compounds including esterified lipids, amino acids, sugars, steroids, and essential oils. The principles underlying these separations are identical to those described in the chapter on separation methods (section 3.6.1). However, the selection of a method of component detection is much more critical, as two important requirements must be fulfilled. First, the system must be capable of continuous operation in order to facilitate successive analyses, and secondly, the detector must give a quantitative response over a wide range of concentrations. In the case of amino-acid analysis these design features have been achieved by the use of a colorimetric assay system involving a flow-cell (see section 5.1.5).

Other successful techniques include the continuous sub-sampling of the column eluate by collection of a thin film of liquid on a moving wire. This sub-sample is transported to an oven where the solvent is evaporated. The solute is then oxidized to carbon dioxide and water. The carbon dioxide is finally reduced to methane, which is measured with a flame ionization detector. It is claimed that this gives a linear response range of 1000:1 using either organic or aqueous mobile phases.

4.3. Bioassay

Tests involving the measurement of the response of living material to an exogenous supply of a particular compound, or group of compounds, are referred to as *bioassay* procedures. These techniques involve the use of whole organisms or excised tissues. The response which is observed may be of the positive/negative type, or it may be a graded effect which can be related directly to the amount of a substance in the sample. With the positive/negative type of response, quantitative estimates are obtained by using a range of sample dilutions and noting the greatest dilution which will induce the effect.

Although different bioassays involve the use of widely differing test materials, certain precautions and design considerations are of general application. Of great importance is the provision of *uniform* biological material. In the case of feeding tests the sex, age, and strain of the animals must be standardized. Plants are usually selected by visual appraisal, to ensure uniformity of size and stage of development, and excised plant organs are often obtained from carefully grown specimens raised under controlled conditions of light intensity, photoperiod, and temperature. In microbiological tests pure cultures are usually used, although there are certain exceptions to this rule. Insects are widely used in the bioassay of pesticides but this particular application is outside the scope of this text. Their use in the evaluation of plant proteins for animal feeding has been attempted but these methods have not been widely adopted. The use of the human in bioassay has one unique application and that is in the assessment of sensory stimuli induced by foodstuffs. Such *organoleptic* methods involve the subjective evaluation of taste, smell, and texture by groups of subjects given samples to examine.

4.3.1. *Tests based on the response of plants or microorganisms*

Bioassays involving the use of higher plants are of great importance in the study of plant growth compounds and many of these techniques are discussed in section 5.7. Both inhibitory and stimulatory compounds can be examined and interactions and synergisms can be evaluated. Although novel systems may be required in a search for hitherto uncharacterized compounds, several of the best-known tests probably account for a large proportion of all bioassays carried out with higher plant materials. The *Avena* coleoptile growth tests, using excised portions of oat seedlings, are among the most widely used procedures. A brief discussion of these particular techniques will serve to illustrate a number of important points relating to bioassays in general.

The conditions of growth of seedlings must be controlled prior to the start of the test to ensure uniform material of an appropriate morphological and physiological condition. The seeds are germinated on a moist support medium, in darkness. After a defined time interval the seedlings are exposed to red light for a fixed period. This treatment suppresses the elongation of the first internode, a prerequisite for the test. After a further period of growth in the dark, sections of coleoptiles are collected from plants within a certain size range and exposed to the sample solutions. During this period of exposure the temperature is closely controlled. Standard solutions containing known amounts of growth regulator are also assayed. Thus in common with other bioassays, the *Avena* tests demand that certain important precautions be taken. These include:

(a) the provision of uniform starting material in a suitable physiological state to respond to the compounds being assayed;
(b) the maintenance of controlled conditions during the assay;
(c) the calibration of the system with standards.

It is usual to find a linear relationship between the logarithm of the concentration of a compound and the biological response which it elicits, at least within certain limits. Control solutions should thus be prepared by a series of tenfold dilutions and a calibration graph drawn. The concentration of samples can then be obtained from the graph.

When the chemical identities of the compounds being examined are uncertain, care must be exercised in the presentation and interpretation of results. For example, compounds which behave as auxins in the *Avene* test may not exhibit properties identical to those of the naturally occurring auxins when some other aspect of plant development is considered. For this reason, special tests may be required in certain studies. An example is the need for complex and specific test systems in studies of the influence of auxins on fruit-set. In many of these methods only a certain selected aspect of growth is measured and this increases the specificity of the procedure.

In some cases the effects produced by a compound or particular type of sample are not well documented. This situation arises in the case of complex materials, such as *polluting effluents* which are suspected as being inhibitory to crop growth. In such cases the usual approach is to carry out initial tests in which the rate of dry-matter accumulation of test plants is measured over a fairly long period. More specific studies can then be considered.

The use of *microorganisms* in bioassay usually involves the measurement of the growth of a bacterial population. Mutant bacteria are easy to isolate and the establishment of a new mutant population is fairly straightforward. An *autotrophic* mutant with a particular biochemical lesion can be used to estimate the compound which, if supplied as a nutrient, will overcome the genetic deficiency and allow the population to increase. Before the advent of automated ion-exchange chromatography, many amino acids were estimated in this way. Because of the special microbiological facilities required these techniques are often unsuited to use in biochemical laboratories.

Microorganisms have also been used to predict the nutritive value of proteins. *Streptococcus zymogenes* and *Tetrahymena pyriformis* are said to exhibit similar amino-acid requirements to those of non-ruminant animals and their rate of growth, on either hydrolysates or intact proteins, has been used as an indication of '*protein quality*'.

4.3.2. *Animal feeding tests*

Although the design and conduct of feeding tests is of greatest interest to specialists in other branches of agriculture and agricultural research, the crop scientist may occasionally encounter the need to undertake or interpret tests of this type. Many feeding experiments are exercises in animal nutrition or production and need not be considered here. However, when the test involves the verification of a prediction obtained by chemical analysis, the mutual co-operation of workers in the plant and animal sciences is required. A good example occurs in the evaluation of crop varieties as sources of dietary protein.

Predictions of the nutritive value of proteins can be obtained by amino-acid analysis. By identifying those amino acids which are likely to limit '*protein quality*' (sections 5.1.6 and 5.2.2), it is possible to screen varieties, or new crops or their products, in order to obtain a comparative estimate of potential feeding value. Eventually these predictions must be tested in animal feeding experiments. The usual laboratory animals for this purpose are the rat and the chick. Less frequently other small animals such as mice are used. Attempts have also been made to use insect larvae whose amino-acid requirements resemble those of man or domestic animals. The larva of the yellow mealworm, *Tenebrio molitor*, has been studied extensively in this context (Davis and Sosulski 1972). The main attraction of this approach is that minute amounts of material can be examined. This is a great advantage in plant breeding programmes where often only small samples can be made available for analysis. Other advantages include the small requirement for laboratory facilities and the ease of statistical replication. However, these techniques have not been widely adopted and many nutritionists consider them unsatisfactory because of the differences which exist between the digestive physiology of insects and domestic animals.

One of the most widely used indices of protein quality is the *protein efficiency ratio* (PER) which is determined by measurement of the growth of the rat (McDonald, Edwards, and Greenhalgh 1973):

$$\text{PER} = \frac{\text{gain in body weight (g)}}{\text{protein consumed (g)}}.$$

Young male rats are chosen for the assay and their increase in body weight is measured over a period of four weeks. The test protein is supplied as 10 per cent of the diet which is provided *ad libitum*. When a particular essential amino acid is present in the sample in an amount which falls short of the full dietary requirement, the rats are unable to make use of all the protein they consume and growth is restricted. This results in a low PER value. Many other tests with laboratory rats have been devised and each one has a set of carefully defined design features which must be closely followed. The chick is used in one popular test which gives an estimate of the gross protein value (GPV). This quantity is determined by comparing the increase in live-weight gain produced by adding 3 per cent of the sample protein to a basal diet, with the increase produced by 3 per cent of a standard, high quality, protein. This standard is the milk protein, casein. Other tests are available but they often involve measurement of nitrogen intake and excretion and involve special cages and equipment.

4.3.3. *Organoleptic Methods*

Many foodstuffs of plant origin are valued for their aesthetic qualities and in recent years scientists have become increasingly aware of the importance of sensory stimuli in the determination of eating habits. Little is gained if new crop products of high nutritive value are introduced, only to be ignored on the grounds of aesthetic appeal. These considerations apply to both low- and high-cost products. In the case of plants grown primarily for their flavouring qualities, attention to the improvement of these properties is an essential component of any programme of genetic improvement, even if the primary aim is greater yield or disease resistance.

The chemistry of flavours is extremely complex and the aesthetic appeal of a foodstuff is also greatly influenced by texture and other properties. Thus the subjective assessment of the sensory appeal of foods is often the preferred approach. The techniques involved are referred to as *organoleptic* methods. Even when attempts are made to characterize flavours in terms of the chemical compounds involved, it is often necessary to run organoleptic assays in parallel. A frequently used method of comparing two samples is the *triangular test*. Samples are grouped together in threes, each group containing one portion of the first sample and two of the second, or vice versa. Each member of a panel of judges is then asked to identify the odd sample. The number of correct identifications is then compared with the number which would occur by chance. The judges may be selected from a larger group of volunteers because of their ability to distinguish flavours or textures, or they may represent a random sample of the population likely to experience the foodstuff under examination. In some cases special precautions are needed to ensure that the basis of the panel's decision is the parameter under consideration. For example, where differences in colour occur, the tasting can be carried out under coloured lights. This prevents an unconscious bias due to the perception of colour preferences.

4.4. Miscellaneous methods

In the previous sections of this chapter, attention has been given to major procedures of wide application. In addition to these techniques there exist many highly specialized methods which, although of less general use, are essential in certain fields of study. Although a full survey of such methods is not possible in this text, three interesting and relatively new groups of methods are outlined below.

4.4.1. *Enzymatic analysis*

Methods based on the measurement of enzyme activity have been known for many years but recently their use has become more widespread due to the ready commercial availability of highly purified enzyme preparations. These assays are based on the measurement of the rate of an enzyme catalysed reaction under conditions in which the substrate is the rate limiting component of the system. Reactions which are coupled to the utilization of NAD^+ or $NADP^+$ are particularly amenable to study as the absorption of light at 340 nm varies according to the concentration of the cofactor. For example, the amino acid L-alanine can be determined with the enzyme L-*alanine dehydrogenase* in the presence of NAD. When the compound under examination is not subject to such a convenient oxidation/reduction reaction, a '*coupled*' assay system can be used, as in the case of glucose.

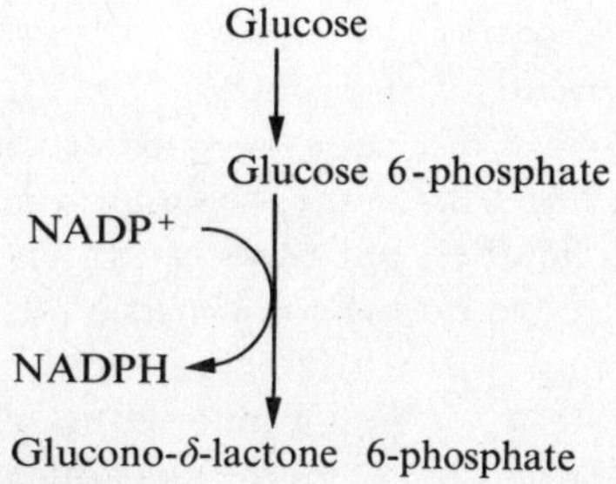

The two enzymes required in this assay are *hexokinase* and *glucose-6-phosphate dehydrogenase*. As the reaction proceeds, NADPH is produced and the rate of production is followed spectrophotometrically at 340 nm. If the reaction mixture in the cuvette is formulated in a fashion which ensures that the initial rate depends

TABLE 4.2

Examples of enzymatic assays involving NAD^+ or $NADP^+$

Compound	Enzyme(s)	Cofactor(s)
Fructose	Hexokinase	NADP
	Glucose 6-phosphase dehydrogenase	ATP
	Phosphoglucoisomerase	
Galactose	β-Galactose dehydrogenase	NAD
Sucrose	β-Fructosidase	NADP
	Hexokinase	ATP
	Glucose 6-phosphate dehydrogenase	
Malic acid	Malate dehydrogenase	NAD
Isocitric acid	Isocitrate dehydrogenase	NADP
Aspartic acid	Glutamate- oxaloacetate transaminase	NADH
	Malate dehydrogenase	

on the concentration of added substrate, then the rate of NADPH production is proportional to the concentration of glucose in the sample added to the mixture. Many other similar enzymatic tests are available and they are particularly well suited to the determination of carbohydrates and organic acids. Some examples are listed in Table 4.2.

4.4.2. *Automated systems of analysis*

Automated procedures for routine analysis have been developed mainly as a result of the efforts of biochemists working in clinical laboratories. However, the basic techniques and the general philosophy of the approach have been introduced into many agricultural laboratories, particularly those concerned with the analysis of materials for farm advisory purposes. The methods chosen for automation must consist of a series of steps which can be linked in a flow system. Procedures involving the separation of solvent phases present technical difficulties, but these can often be solved. Many automated assays depend on the colorimetric estimation of a reaction product occurring between the compound of interest and an added reagent. Peristaltic pumps, connected to glass reaction coils by narrow gauge plastic tubing, are the usual means of reagent and sample transfer. The period of colour development can be adjusted

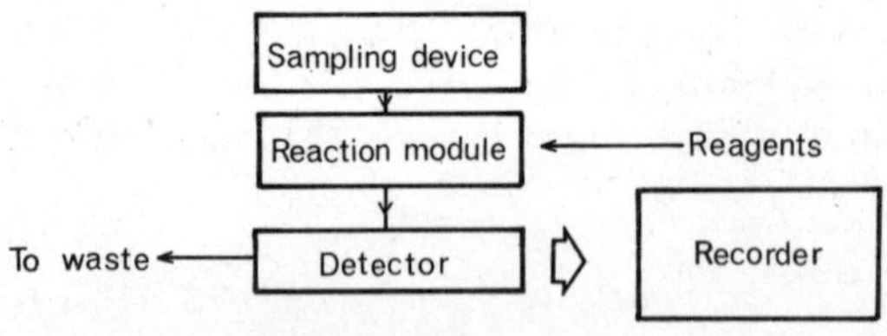

FIG. 4.7. Flow diagram of automated assay.

by means of the flow rate and by the length of the coil. Temperature control is obtained by the use of oil or water baths. Although many different assays have now been developed as automated techniques, they all have common features. In particular, the physical arrangement of the apparatus involves the linking of a sampling device, a reaction module and a detector (Fig. 4.7).

Usually the output of the detector is fed to a recorder which provides for a permanent statement of the results. In order to make the best possible use of such equipment it is important that the large number of data produced can be processed by the laboratory staff at an equivalent rate. For this purpose, the use of programmable calculators, computers, and chart reading devices may be required (see Chapter 6).

An example of the application of automated methods in crop science is the estimation of lysine in protein hydrolysates. This technique involves the enzymatic decarboxylation of lysine and the subsequent colorimetric estimation of the liberated carbon dioxide. The lysine sample is first neutralized and then mixed with the enzyme *lysine decarboxylase*. The solution is buffered at pH 6·5. The reaction mixture is pumped through a glass coil maintained at 37 °C to promote the reaction. After this incubation period the mixture is brought into contact with a carbonate–phenolphthalein indicator. The extent of the decolorization is then measured with a colorimeter incorporating a *flow-cell*.

4.4.3. *Ion-selective electrodes*

Although the glass electrode used in pH measurement is familiar to most workers in biology, there are also other electrodes which respond specifically to individual ionic species. These electrodes are ideal for the rapid, direct measurement of a number of ions including nitrate, chloride, fluoride, and potassium ions. In the conventional glass electrode, a potential is developed across a glass membrane which separates the sample solution from the internal filling solution. In one of the most common types of ion-selective electrodes, the *liquid membrane* type, a similar potential is developed across a layer of liquid ion-exchanger. This layer is formed on an inert porous material and it is selective for the chosen ionic species. The potential developed by the electrode depends on the concentration of this species in the sample solution and this allows for direct analysis.

An important application is the estimation of *nitrate* in various plant materials, including leaf and root tissues. Measurements can also be made on other samples such as exudates formed by different parts of the plant.

REFERENCES

Books

Burchfield, H. P., and Storrs, E. E. (1962). *Biochemical applications of gas chromatography*. Academic Press, London and New York.

McDonald, P., Edwards, R. A., and Greenhalgh, J. F. D. (1973). *Animal nutrition* (2nd ed). Oliver and Boyd, Edinburgh.

Research papers

DAVIS, G. R. F., SOSULSKI, F. W. (1972). Use of the larvae of *Tenebrio molitor* L. to determine nutritional value of proteins in six defatted oilseed meals. *Archs int. Physiol. Biochim.* **80**, 501.

5. Components of plant tissues

5.1. Amino acids

AMINO acids occur in all living plant cells both as constituents of protein and as free acids or amides. More than 200 naturally occurring amino acids have been characterized and many of these have been found in higher plants. Around 20 different amino acids are known to be incorporated into protein and these compounds are thus found universally in living cytoplasm. In general the other amino acids have a much narrower distribution. However, plant species which are able to synthesize unusual amino acids often accumulate these compounds in large amounts. This is particularly true when synthesis occurs in the developing seed. Frequently this part of the plant has been identified as a rich source of certain less common amino acids.

5.1.1. *Classification*

The general formula for an amino acid shows that, with one exception, all amino acids can exist in two stereochemical configurations due to the presence of an asymmetric α-carbon atom.

$$\begin{array}{c} NH_2 \\ | \\ R - C_\alpha - CO_2H \\ | \\ H \end{array}$$

The one exception is the simplest amino acid, glycine, in which the R group is hydrogen.

$$\text{Glycine} \quad \begin{array}{c} NH_2 \\ | \\ H—C—CO_2H \\ | \\ H \end{array}$$

The common naturally occurring amino acids found in plant tissues all have the L-configuration, which means that they resemble L-glyceraldehyde in their arrangement of functional groups.

A convenient method of classifying amino acids is to consider the nature of the R group.

Aliphatic amino acids with hydrophobic R groups. Methionine is unique in this group in possessing a sulphur atom. Although proline has been included it is in fact an *imino* acid. However, it is usually considered in conjunction with the true amino acids due to its wide occurrence in protein.

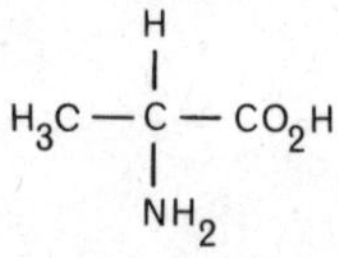

Alanine

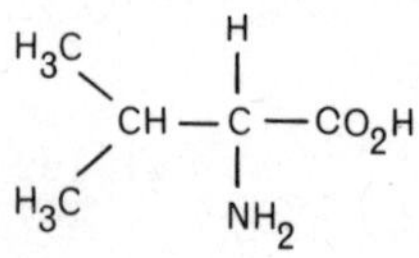

Valine

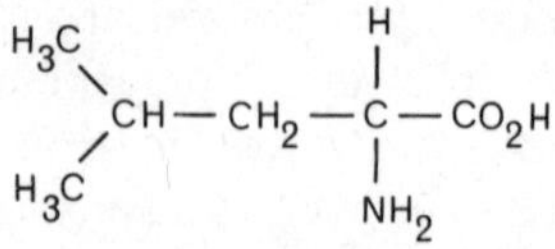

Leucine

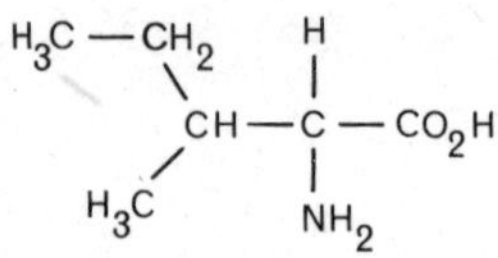

Isoleucine

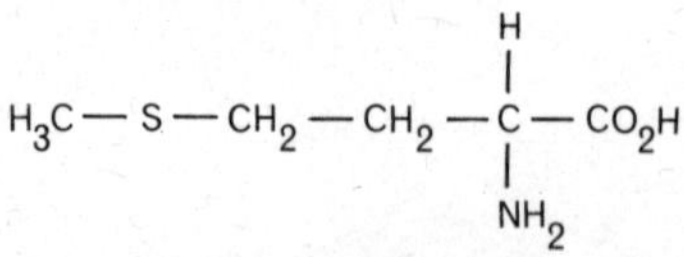

Methionine

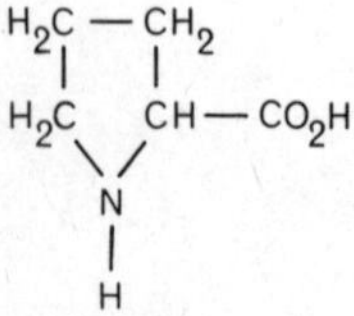

Proline

Aromatic amino acids with hydrophobic R groups. Both the compounds shown, tryptophan and phenylalanine, are important constituents of protein and in addition serve other important roles in cellular metabolism. Tryptophan is notable as the precursor of the important plant hormone, indole-3-acetic acid (IAA).

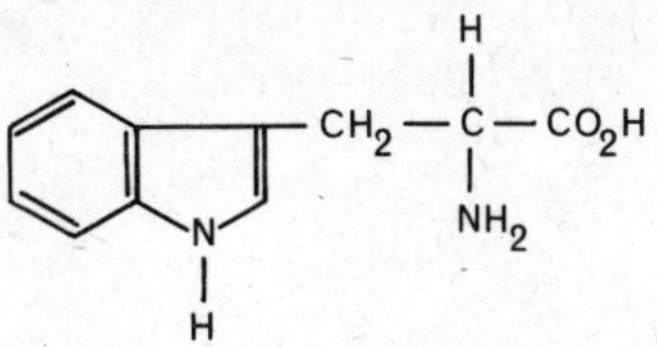

Tryptophan

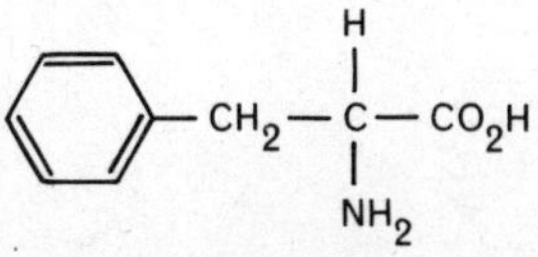

Phenylalanine

Amino acids with polar side groups fall into one of two classes. In the first the R substituents are charged at physiological pHs, either positively (basic) or negatively (acidic) and, in the second, the substituent groups are uncharged.

Basic amino acids. Each of the basic amino acids makes an important contribution to the unique properties of many individual proteins. Histidine in particular is often found in the region of the 'active' or *catalytic* site of enzyme proteins.

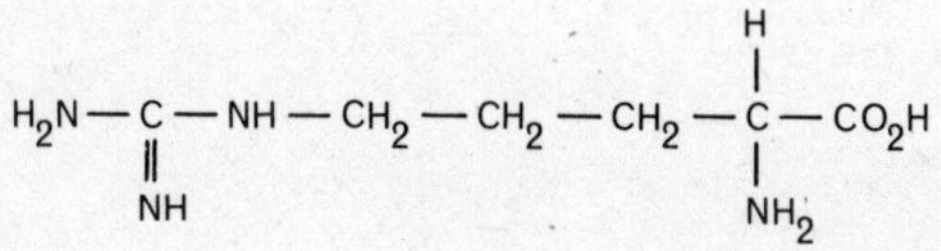

Arginine

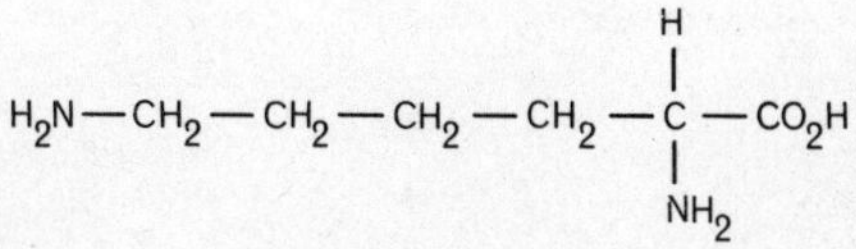

Lysine

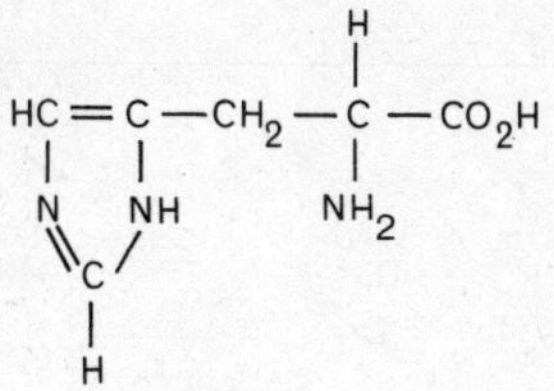

Histidine

Acidic amino acids. Glutamic acid is a major constituent of the *prolamin* fraction of cereal protein preparations, while aspartic acid occupies an important position as a precursor of several other amino compounds, including lysine.

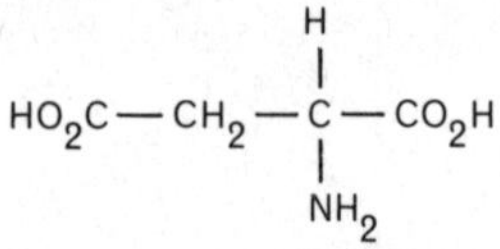

Aspartic acid

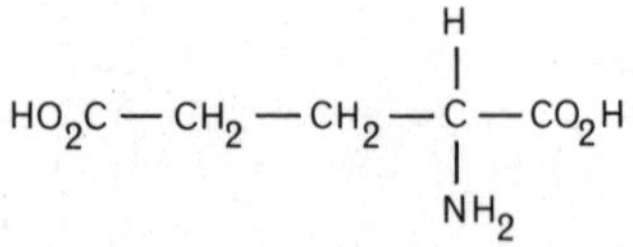

Glutamic acid

Other amino acids with polar R group.

(a) Containing hydroxyl groups;

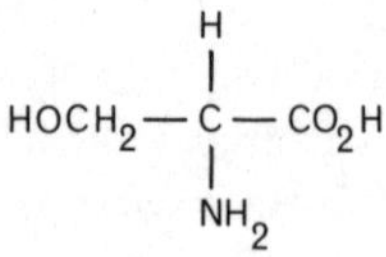

Serine

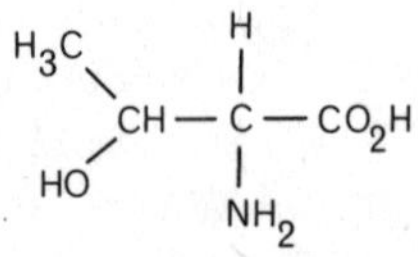

Threonine

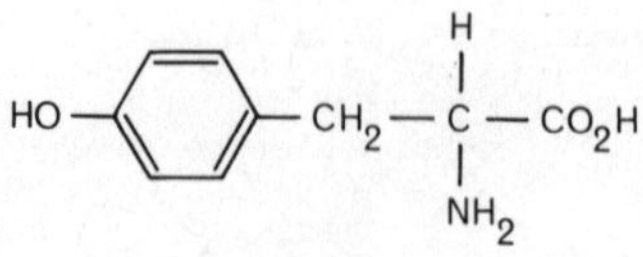

Tyrosine

(b) Amides;

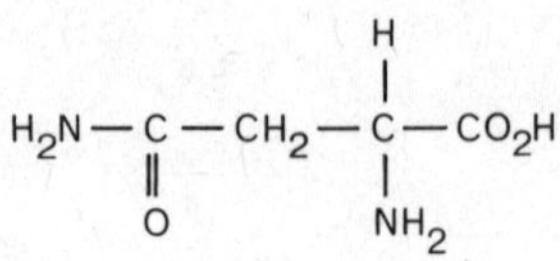

Asparagine

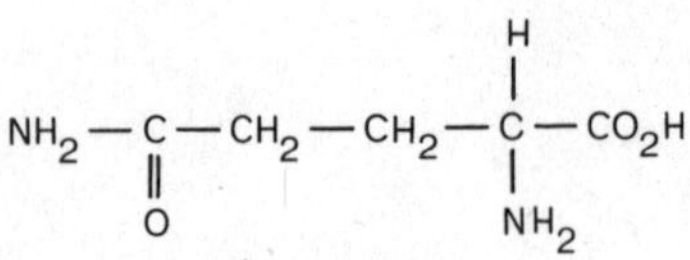

Glutamine

In most plant tissues the amides, asparagine and glutamine, are extremely important components of the pool of free amino acids. They occur in exceptionally large amounts in certain specific plants, asparagine being found in high concentration in asparagus (*Asparagus officianalis*) whence its name, while glutamine accumulates in the fleshy storage tissue of the beetroot (*Beta vulgaris*). These compounds are also major components of the fluid moving through the xylem vessels and they serve as a means of transporting organic nitrogen from the root system to the leaves and to the apical regions of the plant. Often up to two thirds of the nitrogen exported by the root is transferred in this way.

(c) Containing sulphur.

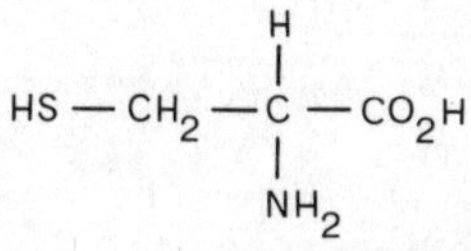

Cysteine

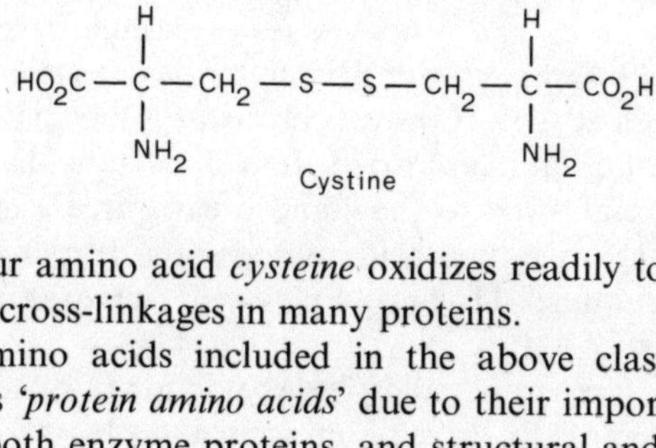

Cystine

The sulphur amino acid *cysteine* oxidizes readily to give *cystine*, which forms cross-linkages in many proteins.

All the amino acids included in the above classification are referred to as '*protein amino acids*' due to their importance as constituents of both enzyme proteins, and structural and storage proteins. Many other amino acids occur in plants, some confined to a single genus. An example which occurs only in *Vicia* species is *γ-hydroxyarginine*. Twelve other non-protein amino compounds

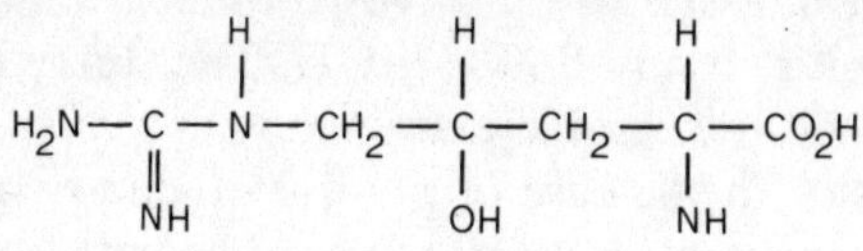

γ-Hydroxyarginine

occur in this genus and the distribution of these substances has been used in taxonomic studies.

Unusual amino acids are also found in the seeds of many other legumes. Some of these compounds have been studied in considerable detail and a recent survey has shown that they serve a protective function, having either toxic or repellent properties towards insect pests. An example is the compound β-*hydroxy*-γ-*methylglutamic acid* which constitutes approximately 4 per cent of the dry matter of seeds of *Gymnocladus dioicus* (kentucky coffee tree).

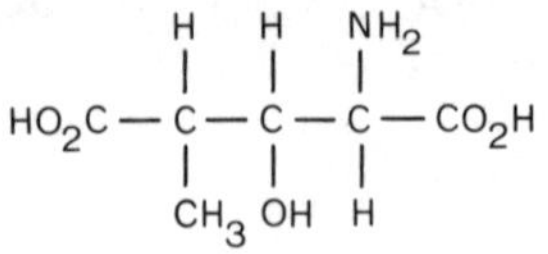

β-Hydroxy-γ-methylglutamic acid

5.1.2. *Extraction of free amino acids*

The first step in the analysis of the free-amino-acid components of plant material is to establish an appropriate sampling procedure. Ideally leaves, fruits and other tissues containing large amounts of water should be sampled immediately prior to extraction, or alternatively, frozen at −18 °C as quickly as possible and then stored at this temperature, or freeze dried. It is important that deep-frozen material is not allowed to thaw and then re-freeze as this process leads to changes in amino-acid composition. In particular there is a loss of glutamine. Drying at elevated temperatures is not recommended.

Extracts of free amino acids may be obtained by macerating the tissue in an electrically driven homogenizer or by the use of a pestle and mortar, with a little acid-washed sand as an aid to cell disruption. A suitable solvent is aqueous ethanol, so adjusted to give a concentration of 75–80% ethanol in the final homogenate. A typical procedure for leaf tissue would involve the extraction of 5·0 g of fresh material with 50 ml of 80% EtOH. The slurry is then filtered through paper and the residue extracted with two further portions (50 ml) of solvent. In the case of photosynthetic tissue the chlorophylls and other pigments must be removed by partition into organic solvent. To achieve this end, the ethanolic extract is reduced to dryness *in vacuo* at a temperature not exceeding 40 °C and the residue shaken with a mixture of water (25 ml) and light petroleum (b.p. 40–60 °C). The upper layer is discarded and the procedure is repeated twice. The final pigment-free aqueous phase, together with any undissolved solids, is reduced to dryness and the amino acids dissolved in 10 ml 0·1 M HCl.

Similar procedures may be applied to air-dried substances such as cereals and grain products. Usually however, greater importance is attached to the total amino-acid content of such materials.

Occasionally it is useful to examine liquid or semi-liquid preparations such as exudates taken from particular plant tissues. The procedure in such cases is to add to the sample (say 1·0 ml) a quantity of absolute ethanol (20·0 ml) sufficient to precipitate any proteinaceous constituents. Subsequently the extract can be filtered or centrifuged and the liquid phase taken directly for thin layer chromatography (t.l.c.). Alternatively the ethanol may be removed by rotary evaporation, the solids re-dissolved in 0·1 M HCl and the solution examined by column chromatography.

5.1.3. *Protein hydrolysis*

Usually only a small proportion, say 5–10 per cent, of the total amino-acid residues are found in the uncombined form. In order to estimate the total amino-acid content of plant tissues it is first necessary to degrade the proteins by hydrolysis. If a separate estimate of free amino acids is required, the plant material should preferably be stored in the deep-freeze prior to analysis, in order to minimize amino-acid interconversions and increments to the free-amino-acid pool arising as a result of proteolysis. Free amino acids can be removed with 80 per cent EtOH and the residue retained for protein hydrolysis. When the requirement is for an estimate of total amino acids, it is usually regarded as satisfactory to dry material in a forced draught oven at around 80°C.

Three possible means of protein hydrolysis have been employed in amino-acid studies but the technique which is used in most laboratories is acid hydrolysis. *Enzymatic methods* often give incomplete breakdown of proteins and the use of proteolytic enzymes is usually confined to laboratories with a particular interest in the properties of plant proteins in relation to animal physiology and nutrition, in which case the enzyme methods may reveal important characteristics of proteins likely to influence their nutritive value.

UNIV. COLL. N.W. BANGOR LIBRARY

Hydrolysis with alkali is employed when estimates of tryptophan are required as this compound is degraded in acid hydrolysates.

A typical procedure for the *acid hydrolysis* of dried plant material, such as milled cereal grains, is outlined below.

An accurately weighed sample of approximately 250 mg is transferred to a 500 ml flask fitted with a water-cooled reflux condenser. To the flask is added 420 ml of 6N HCl and the mixture is flushed with nitrogen. The large volume of acid minimizes condensation of the ε-NH_2 group of lysine with carbohydrate. Hydrolysis is carried out for 20 hours at 110 °C. Nitrogen is bubbled through the reaction mixture and this is continued for 30 minutes following the removal of heat. The hydrolysate is filtered through paper and made up to 500 ml with distilled water. An aliquot is then reduced to dryness (40 °C) and the amino acids dissolved in 0·1 M HCl (20 ml) and stored at −18°C.

5.1.4. *Thin layer chromatography*

One of the most effective methods of separating amino acids is t.l.c. (Stahl 1965). This technique is extremely sensitive, 1 μg of amino acid producing a deeply staining spot on the developed plate, while smaller amounts are easily detectable. The method is also fairly quick to perform, particularly if development in only one dimension is required. Another advantage is the low cost of the apparatus. Despite these attractions, one major disadvantage exists and that is the difficulty of quantification. This is usually accomplished by eluting the amino acids and carrying out a colorimetric assay. However, a much more convenient method for quantitative work is column chromatography, for which there are now rapid automated systems commercially available.

The main use of t.l.c. in amino-acid studies is for qualitative and preparative work, particularly if the identity of compounds is in question. The support material is usually silica gel and the coated plates are dried at room temperature overnight. For two-dimensional separations the following solvents are recommended, used in the order indicated.

Solvent 1: n-butanol:acetic acid:water (80:20:20 v/v)

Solvent 2: phenol:water (75:25 w/v)

It is essential that the phenolic solvent is used for the second dimension as it as almost impossible to evaporate from the plate and any 'carry-over' of solvent interferes with subsequent development. A suitable sample volume is 1 μl, using either 0·1 M HCl or 80 per cent EtOH as the solvent. Preparative plates will require larger sample loadings.

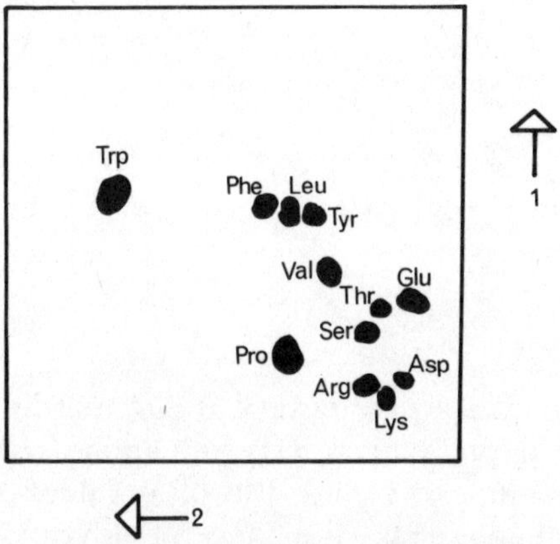

FIG. 5.1. Thin-layer chromatogram of amino acids in an hydrolysate of barley prolamin, developed in two dimensions with (1) n-butanol:acetic acid: water (80:20:20 v/v) and (2) phenol:water (75:25 w/v). The sample was applied to the silica gel layer at the bottom right hand corner of the plate. After development the amino acids were detected with ninhydrin.

After development the plates are dried either at room temperature, in the case of preparative work involving heat-labile compounds, or at elevated temperature for qualitative studies, and then sprayed with *ninhydrin reagent* (Fig. 5.1). Coloured spots appear following 3 minutes heating at 105 °C. Pre-filled aerosol packs of ninhydrin reagent are now available and generally these are preferred for their convenience and excellent spray pattern.

Most of the amino acids give a purple coloration with ninhydrin, with a simultaneous liberation of carbon dioxide. Proline is unusual in giving a yellow product which makes identification of this compound particularly easy. With experience, it is also possible for the analyst to identify other amino acids by their colour reaction with ninhydrin. For example, glutamic acid gives a blue spot on thin layer plates. Because of the variations which occur in colour assessment

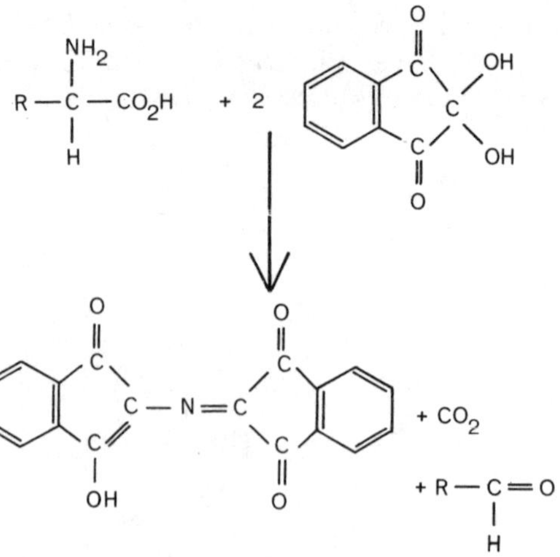

Coloured product

from one individual to the next, it is essential for each worker to carry out careful preliminary tests with amino-acid standards.

Some analytical texts include lists of R_f values (see section 3.6.3) for different amino acids, in a range of solvents. This practice is frequently misleading as the small differences in technique which inevitably occur from one laboratory to another can cause quite appreciable variations in R_f values. It is thus important that a complete range of standards is examined whenever a separation is being undertaken for the first time.

5.1.5. *Automated column chromatography*

This technique, originally conceived by Moore and Stein in the early 1950s, depends on the use of an ion-exchange resin from which the different amino acids are eluted with a range of buffered solutions (Moore and Stein 1954). Recent developments include greatly improved grades of resin, automatic loading procedures employing sample cartridges, higher operating pressures, and shorter and narrower columns. These improvements in design have resulted in a decreased time period for the separation, results being available for protein hydrolysates within two hours of sample application. However, some commercially available instruments are so critical in their operation that this increase in speed has been achieved only at the expense of a loss in reliability.

Usually two separate columns are used, one for the basic amino

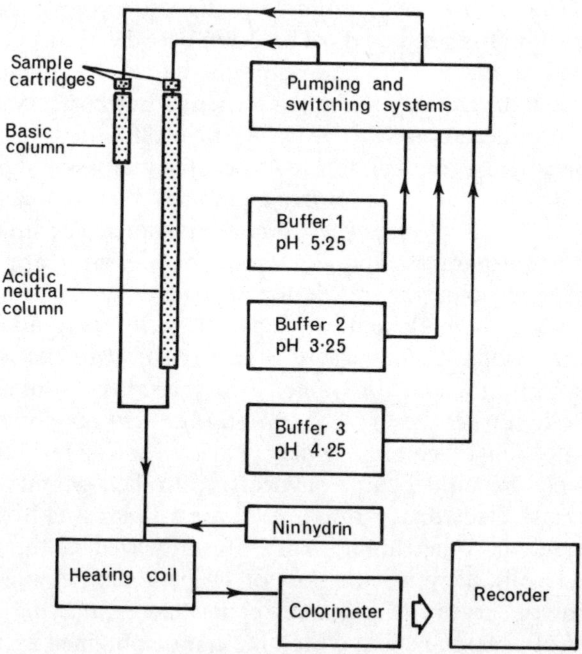

FIG. 5.2. Automated system for column chromatography of amino acids.

acids, lysine, histidine, and arginine; and a longer column for the non-polar, acidic, and uncharged polar amino acids (Fig. 5.2). Samples are applied to each column, either by means of direct pumping of the solution from a holding-coil of plastic tubing, or by means of a small resin-filled cartridge containing the amino-acid mixture. In the latter instance, the cartridge is simply inserted into the buffer flow immediately above the top of the column. This method of loading and the general arrangement of the anlytical system is shown diagrammatically in Fig. 5.2. Three separate buffers are used: the first, at pH 5·25, is used to elute the basic amino acids from the short column and the other two (pH 3·25 and pH 4·25) are used sequentially to elute the acidic and neutral amino acids from the second column. From the column outlets the amino acids are swept towards the colorimeter. Before reaching the optical system the eluate is mixed with ninhydrin reagent and the colour reaction is allowed to proceed as the mixture passes through a glass coil maintained at 95 °C. The optical density is measured at two wavelengths, one appropriate to the yellow product arising from the reaction between proline and ninhydrin, the other for measurement of the remaining amino acids. The output to the recorder is continuous as the colorimeter operates on the 'flow-cell' principle.

The latest development is an alternative detection system involving fluorimetric assay. This new apparatus offers greatly increased sensitivity and will no doubt become increasingly popular.

The separation of extracts of free amino acids rather than protein hydrolysates often involves several complications. The first is that if the amides, asparagine and glutamine, are present in appreciable quantities, the degree of resolution provided by the usual two-column system is inadequate to separate these compounds from other amino acids. Usually there is an overlap with the hydroxy-amino acids, serine and threonine. Two solutions to this problem are available. Either the two-column arrangement may be replaced by a single, longer column which provides better resolution, or alternatively the amides can be hydrolysed to their parent acids. In the latter case the hydroxy-amino acids are estimated in hydrolysed samples and the other amino acids are measured in the original extract. Usually there is some loss of serine and threonine during the hydrolysis treatment and a correction factor must be applied. Estimates of asparagine and glutamine can be obtained by measuring the increments in aspartic acid and glutamic acid respectively, following hydrolysis. The conditions of hydrolysis are usually 1 M HCl for 3 hours, at 100 °C. Care should be taken to exclude oxygen from the reaction mixture by flushing with nitrogen.

A second difficulty in working with free amino acids is that often a far greater range of compounds is present. This may lead to problems of resolution and also of identification. In addition to non-protein amino acids, other ninhydrin positive compounds may occur, including amino sugars and peptides. One important peptide frequently observed is *glutathione*. This molecule, which consists of three amino-acid residues linked by peptide bonds, is more correctly named *γ-glutamylcysteinylglycine*.

If an unidentified peak fails to correspond with any reference

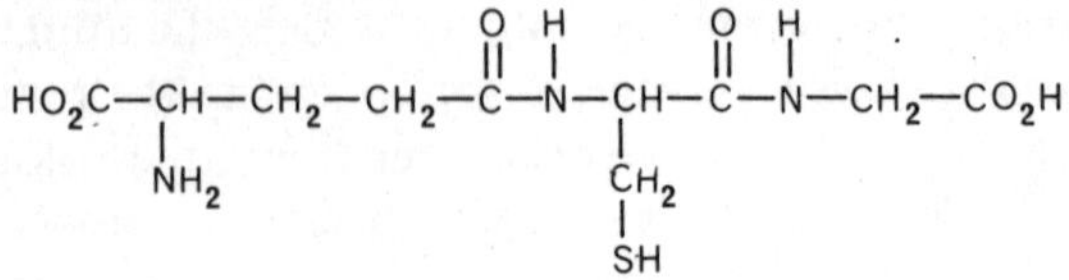

Glutathione

compound, then a further examination must be undertaken by some other means such as t.l.c.

5.1.6. *Essential amino acids*

In contrast to higher plants and microorganisms, animals lack the ability to synthesize certain of the protein amino acids. These compounds are referred to as *essential amino acids* due to the fact that they must be supplied in animal diets. Although different animals vary in their requirements, a supply of all, or most, of the following is usually obligatory: lysine, methionine, threonine, tryptophan, valine, arginine, histidine, phenylalanine, leucine, and isoleucine (McDonald, Edwards, and Greenhalgh, 1973). Ruminants, such as cattle and sheep, are able to utilize protein synthesized by the rumen microorganisms. For this reason, the exact amino-acid content of the diets of these animals is of little consequence, particularly in comparison with non-ruminants such as man, pigs, and poultry.

The major concern of the crop scientist is to ensure that those plants and plant products which are important as sources of protein for non-ruminant animals contain sufficient quantities of the essential amino acids. In recent years increased attention has been focused on the amino-acid content of cereals, as these crops supply a large proportion of the dietary protein of certain sectors of the world population. Cereals are usually deficient in lysine and sometimes in tryptophan and other amino acids. At a number of centres work is now in progress aimed at improving the lysine content of cereal varieties by plant breeding. These attempts at cereal improvement are of major importance, as the inefficient utilization of protein is economically and socially unacceptable in the face of a world food shortage.

Accurate, yet rapid, analytical methods for estimating lysine are vital to plant breeding programmes of this type. Automated column chromatography is widely used, and frequently the time of analysis can be decreased by abandoning the use of a long second column, and simply estimating the basic amino acids on a short column using a single buffer at pH about 5·25.

One special problem exists in the assessment of the ability of crop products to contribute lysine in a form in which it is assimilated by the feeding animal. This is the difficulty of determining the extent to which the lysine is made '*unavailable*' by complexing of the ε-amino group to other chemical residues. The complex is degraded during acid hydrolysis but it is not affected by enzymes in the gastrointestinal tract of animals. Unavailable lysine can be estimated by treating the protein with fluorodinitrobenzene or nitrous acid and then, following hydrolysis, estimating the unmodified lysine residues which remain (Milner and Westgarth 1973).

5.1.7. *Other procedures*

In recent years a considerable advance has been achieved in the analysis of amino acids by *gas–liquid chromatography* (Burchfield and Storrs 1962). However, this technique is not suitable for all amino acids and ion-exchange chromatography is still preferred in most situations. Amino acids are involatile compounds and thus derivatives must be prepared prior to separation in the gas phase. The usual derivatives are either *N–trimethylsilyl esters* or *N-acetyl-n-amyl esters*.

Amino acids can also be estimated by *microbiological methods* and many early analyses of plant proteins were carried out in this way. The basis of many of these procedures is the measurement of the growth of the appropriate *autotrophic* mutant when supplied with a sample of amino acid. For example, the growth of a bacterium lacking the ability to produce valine can be related to the concentration of this amino acid in the sample under examination.

5.2. Proteins

Although all proteins consist of amino-acid residues linked by peptide bonds, their molecular size, shape, physico-chemical properties, and biological role vary considerably from one molecular species to another.

Frequently differences in properties are reflected in different functions within the plant and thus the classification of proteins according to their chemical nature is convenient for the biologist as well as the analyst. Much of the early work on plant proteins was carried out by Osborne whose monograph *The vegetable proteins* (1909) is still occasionally referred to by research workers concerned

The peptide bond

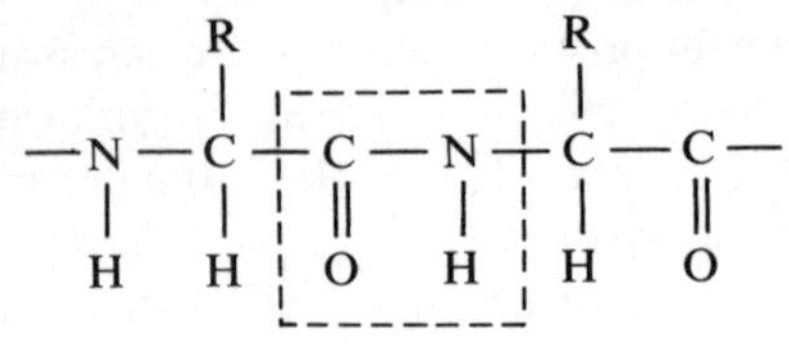

with the separation of seed proteins. Osborne classified proteins according to their solubility properties.

Albumins water soluble;
heat coagulable;
usually enzyme proteins.
Globulins soluble in salt solutions;
precipitated by dialysis;
enzyme or storage proteins.
Prolamins soluble in aqueous ethanol;
precipitated with absolute ethanol;
seed storage proteins.
Glutelins insoluble in neutral aqueous solutions and neutral ethanol;
seed storage proteins.

Proteins acting as *structural components* of cellular membranes are not conveniently included in these four groups and are best considered separately. This comment also applies to the *histones*, proteins which occur in the cell nucleus. Most *enzyme proteins* are soluble in buffered aqueous solutions and their detailed classification relates to their catalytic activity rather than any other characteristic.

In crop science it is sometimes sufficient, and indeed desirable, to simply carry out estimates of the total protein content of such plant materials as cereal grains, roots, tubers, fruits, and leafy products.

5.2.1. *Total 'crude-protein' content*

It is common practice in most laboratories to obtain an approximate estimate of the protein content of crop products by estimating the nitrogen content of samples, and applying the following formula:

$$\text{percentage crude protein} = \text{percentage nitrogen} \times 6{\cdot}25.$$

The conversion factor is derived from the assumption that protein contains 16 per cent nitrogen. In fact different proteins vary in their nitrogen content due to the presence of different proportions of individual amino acids. Furthermore, some nitrogen is contained in nucleic acids, free amino acids, and other classes of compounds such as the alkaloids. Despite these criticisms, the use of crude protein values is an extremely useful convention. A particular advantage is that estimations of nitrogen can be carried out conveniently and with high precision, using the conventional *Kjeldahl* procedure

(Association of Official Agricultural Chemists 1970). This technique involves the digestion of samples with sulphuric acid in the presence of a catalyst, followed by the determination of the ammonia produced. The procedure is not fully sensitive to nitrate, which is only partially reduced. However, the degree of reduction is sufficient to give misleading results for protein, in crops receiving large amounts of nitrogenous fertilizers.

5.2.2 *Storage proteins*

Seed storage proteins are important as reserves of organic nitrogen which are utilized during germination by the developing embryo. Following the imbibition of water by the seed, proteolytic enzymes are responsible for the release of free amino acids which are eventually used for the production of cellular proteins. As the amino-acid requirements of the embryo differ markedly from the amino acids available as a result of proteolysis, other enzymes are essential for the full utilization of the protein reserves. For example, in cereals there are usually larger amounts of proline and glutamic acid than are required by the embryo and thus *transaminase* enzymes are required for the appropriate amino-acid conversions.

As well as their importance in seedling development, storage proteins are of great interest in agriculture due to their presence in many crops destined for human or animal feeding. Most seeds contain at least small amounts of albumin proteins and these are usually of high nutritive value due to their well-balanced amino-acid content. The globulins and glutelins frequently contain slightly less favourable proportions of the different amino acids but nevertheless are assumed to be of reasonable feeding value. The prolamins, however, contain extremely low levels of lysine and as the proportion of the total seed protein represented by prolamin components increases, the over-all amino-acid composition becomes less favourable from the nutritional standpoint. These comments apply mainly to the nutritional requirements of man, pigs, poultry, and other non-ruminant animals (section 5.1.6). In ruminants such as cattle and sheep, dietary amino acids are degraded and re-synthesized by microorganisms in the rumen, and the animals are able to utilize low quality proteins and even certain amounts of non-protein nitrogen such as urea.

Factors which affect the proportions of the different proteins laid down during seed maturation include genotype, fertilizer treatment, and environment. In barley, as in many cereals, high levels of fertilizer nitrogen result in grain of higher protein content but the *proportion* of lysine, the amino acid which usually limits the nutritive value of cereal proteins, decreases with total protein content. Ideally, feed grains would contain around 15 per cent protein and the lysine content would be much higher than in presently available commercial varieties. Steps are now being taken to improve the amino-acid content of cereals by plant breeding. Already improved varieties of maize (opaque-2 and floury-2) are under cultivation. These varieties have higher contents of tryptophan and lysine due to a decreased proportion of the maize prolamin, zein, a fraction extremely deficient in these amino acids.

Because of the current world-wide interest in cereal improvement, the fractionation of seed proteins has recently received renewed interest. The albumins are usually present in only small amounts and they are often grouped with the globulins to form a '*salt-soluble*' fraction. Many components of this fraction are enzymes and thus are not regarded as true storage proteins. A procedure for the isolation of 'salt-soluble', prolamin and glutelin proteins from cereals is outlined schematically in Fig. 5.3 and 5.4.

Interest is also attached to cereal proteins on account of their importance in determining the bread-making properties of wheat flours. Frequently wheat proteins are extracted with dilute acetic acid, a practice which yields a heterogeneous protein preparation referred to as *gluten*. In the presence of water, wheat gluten displays considerable elasticity and this property ensures that gases produced by the yeast fermentation become entrapped in the dough. Gluten, in fact, contains both prolamin components, the *gliadins*, and glutelin proteins, the *glutenins*.

The precise details of any scheme of fractionation must be chosen to suit the particular species in question. The procedures shown in Fig. 5.3 and 5.4 have been applied successfully to oats (Draper 1973) and barley (Folkes and Yemm 1956). It is important to define exactly the conditions employed as the problem of identifying storage proteins unambiguously is considerable, due to the absence of unique properties analogous to the specific activities displayed by enzyme proteins. The final protein preparations should be freeze dried rather than oven dried, in order to avoid oxidative deterioration.

As an illustration of differences in the amino-acid contents of two different seed protein fractions, amino-acid analyses of barley proteins are given in Table 5.1. Characteristically the prolamin fraction contains large amounts of glutamic acid and proline, and only a small amount of lysine.

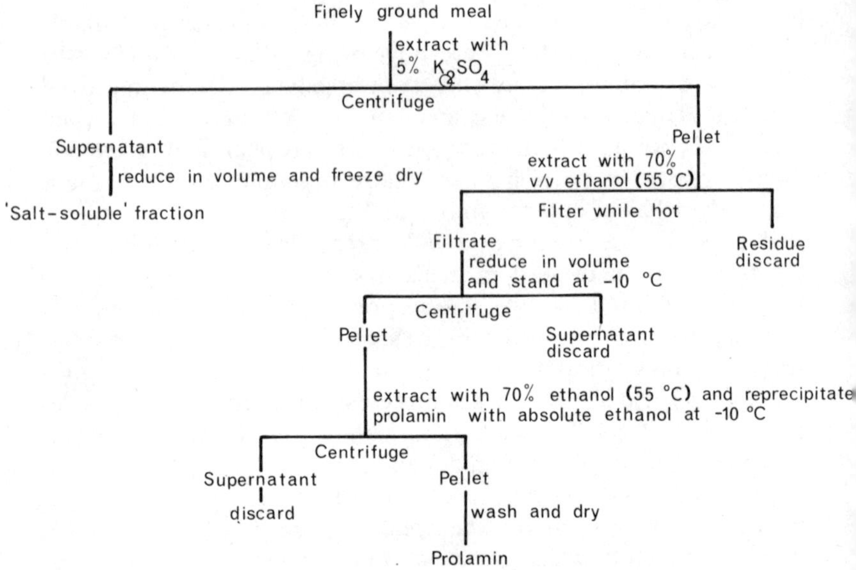

FIG. 5.3. Salt-soluble and prolamin proteins.

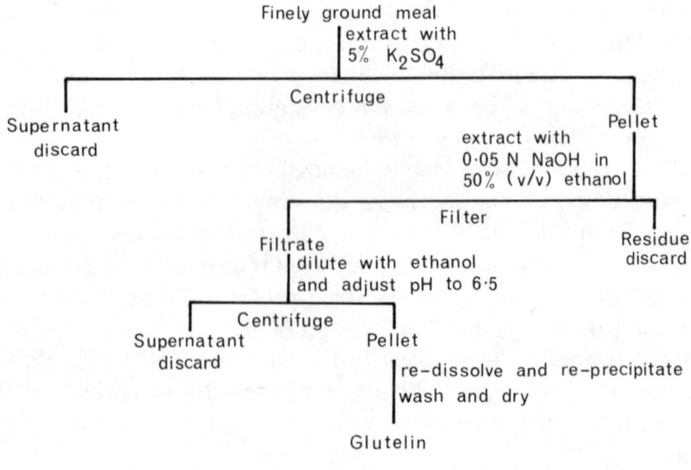

FIG. 5.4 Glutelin proteins.

TABLE 5.1

Amino-acid composition of the salt-soluble and prolamin protein fractions of barley grain (var. Nackta)

	Salt-soluble proteins	(g per 16 g nitrogen)	Prolamin
Lysine	6·41		1·08
Histidine	2·94		1·39
Arginine	7·64		2·24
Aspartic	11·04		2·13
Threonine	4·66		2·28
Serine	4·86		3·69
Glutamic	14·21		36·86
Glycine	6·28		2·23
Alanine	7·44		2·75
Cysteine	3·56		1·45
Valine	5·83		4·25
Methionine	1·41		1·46
Isoleucine	3·73		3·79
Leucine	6·53		6·91
Tyrosine	4·29		3·68
Phenylalanine	4·41		7·09
Proline	5·39		20·57

FIG. 5.5. Polyacrylamide gel electrophoresis of prolamin proteins obtained from oats (Draper 1973). The sample was loaded at the top of the gel.

Protein preparations can often be characterized by electrophoresis or other physico-chemical methods and an example is shown in Fig. 5.5. This diagram illustrates the separation of prolamin proteins from oats using polyacrylamide disc–gel electrophoresis (Draper 1973). The fractionation demonstrates the polydisperse nature of a prolamin preparation which was obtained in the manner outlined in Fig. 5.3.

In legumes and in certain other important groups of plants, the globulin fraction assumes a dominant position as the major storage protein. Some of these globulin preparations form crystals even when a number of different components are present. Extracts are usually prepared in a solution of 1–2 M NaCl using de-fatted meal as the starting material. In soya bean extracts prepared in this way, four major components are observed, differing in sedimentation characteristics. Some species exhibit distinct aggregations of protein in the endosperm known as *protein bodies*. Protein bodies from species of the pea, *Pisum*, contain almost exclusively globulin material.

Globulins are also present in large amounts in members of the *Cucurbitaceae* such as the winter squashes and pumpkins *C. maxima*, which yield crystalline preparations of *cucurbitin* following extraction of seeds with a solution of sodium chloride. Many other plant-globulins have been isolated, characterized, and named (Osborne 1909). Some examples are given in Table 5.2.

TABLE 5.2
Some plant globulins acting as storage proteins in seeds

Protein	Source
Glycinin	*Glycine max* (soybean)
Arachin Conarachin	*Arachis hypogea* (groundnut)
Legumin Vicilin	*Pisum sativum* (green pea)
Edestin	*Cannibis sativa* (hemp)
Avenalin	*Avena sativa* (oats)

5.2.3. *Structural proteins*

Although a certain degree of uncertainty surrounds the exact structural configuration of biological membranes, one common feature of such systems is the presence of protein components in addition to the membrane lipids (Rothfield and Finkelstein 1968). Structural proteins are usually regarded as polypeptides lacking enzyme activity and playing an important role as elements of cellular membranes. This definition must be regarded with a certain degree of caution as it is now known that many enzymes associated with membranes possess activity only in the presence of a lipid component. Thus certain protein fractions previously regarded as devoid of enzyme function are now known to fulfil a catalytic role.

Proteins which are believed to fulfil a structural role have been extracted from a number of plant tissues and studies of the physico-chemical properties of these materials have been carried out. A chloroplast protein can be obtained from lipid-free preparations of chloroplast lamellae by extraction with tris chloride buffer (pH 8·5) containing potassium *deoxycholate* and potassium *cholate*. An alternative detergent is *sodium dodecyl sulphate* in the presence of urea. These techniques have been described in detail by Criddle (1966) who has obtained similar preparations from mitochondria. Other agents which have been used to obtain solutions of structural protein include urea, at high concentration (8M) but in the absence of detergent, ammonium hydroxide (1.2M), and acetic acid.

Because of their sparingly soluble nature, structural proteins are difficult to study in electrophoretic systems and in the ultracentrifuge. However, separations which have been reported in the literature suggest that the molecular weight of both mitochondrial and chloroplast structural proteins is around 25 000. In the case of chloroplasts this value is supported by X-ray studies of chloroplast membranes which indicate the presence of structural repeating units of this size. Gel electrophoresis has been used successfully to demonstrate polydispersity in chloroplast protein preparations.

Structural proteins tend to contain higher proportions of non-polar amino acids than most enzymes and this property undoubtedly favours the formation of hydrophobic bonds with lipid molecules. Structural proteins are also known to bind cytochromes and mitochondrial enzymes with consequent effects on the properties of these molecules.

Despite the considerable number of investigations of 'structural' protein fractions, it is still not clear whether or not many of these materials are really artefacts produced as a result of enzyme denaturation. For this reason, great care should be excercised when attempts are made to interpret or repeat some of the earlier work in this field.

5.2.4. *Enzymes*

Enzymes occur universally in living tissues and frequently a large number of different activities can be detected in suitably prepared extracts. Basic knowledge of the distribution, properties, and function of plant enzymes is now more than sufficient for greater use to be made of enzyme assays in investigations of the growth and development of crop plants and the storage of fruits and other plant materials. Furthermore, in situations where the activity of a single enzyme can be related to some particular aspect of development or to some important property of the final plant product, there exists the possibility of using enzyme techniques as a tool for the selection of improved genetic material.

Unlike the different classes of storage proteins referred to in the previous section, enzymes are classified according to their precise role rather than to their physico-chemical properties. Unfortunately, two conventions exist for the naming of enzyme proteins. Most enzymes possess a *trivial name*. For example, the important plant enzyme responsible for the initial step in nitrate reduction is known as *nitrate reductase*:

$$NAD(P)H + NO_3^- \xrightarrow{\text{NITRATE REDUCTASE}} NAD(P)^+ + NO_2^- + H_2O.$$

However, a more rigorous procedure for naming enzymes is now widely recognized, in which the enzyme is defined by a reference number and by a specific and systematic name. Thus nitrate reductase is renamed:

EC 1.6.6.2 NAD(P)H:nitrate oxidoreductase

The system was devised by the Enzyme Commission (EC) of the International Union of Biochemistry (1961). The first number refers to one of six major groups.

EC 1. *Oxidoreductases*: these enzymes are concerned with the oxidation or reduction of subtrates.

EC 2. *Transferases*: the reactions catalysed by these enzymes involve the transfer of a group from one substrate to another.

EC 3. *Hydrolases*: catalyse the breakdown of a molecule into two products and involve the utilization of water.

EC 4. *Lyases*: these enzymes split molecules without the direct involvement of water in the reaction.

EC 5. *Isomerases*: bring about the rearrangement of atoms in the substrate molecule.

EC 6. *Ligases*: catalyse the combination of two molecules with the utilization of a 'high energy' compound such as ATP.

Thus the reference number for NAD(P)H:nitrate oxidoreductase, EC 1.6.6.2, places the enzyme in Group EC 1, the oxidoreductases. The other numerals refer to further sub-divisions:

1.6—oxidoreductases using NADH or NADPH as donor.

1.6.6—with a nitrogenous group as receptor.

The final numeral is specific to NAD(P)H:nitrate oxidoreductase.

An example from each of the major EC groupings is given in Table 5.3. Included in this list is *isocitrate glyoxalate-lyase*, a key enzyme in the glyoxalate cycle. This pathway, which is found only in plants

TABLE 5.3

Classification and function of some important plant enzymes

Number	Systematic name	Trivial name	Reaction
1.6.6.2	NAD(P)H:nitrate oxidoreductase	Nitrate reductase	NAD(P)H+ nitrate= NAP(P)$^+$ +nitrate+H_2O
2.2.1.1	D-Sedoheptulose 7-phosphate:D-glyceraldehyde 3-phosphate dihydroxyacetone-transferase	Transaldolase	Sedoheptulose 7-phosphate + Glyceraldehyde 3-phosphate = erythrose 4-phosphate + fructose 6-phosphate
3.1.1.3	Glycerol-ester hydrolase	Lipase	Triglyceride + H_2 O = diglyceride + fatty acid
4.1.3.1	Isocitrate glyoxalate-lyase	Isocitrate lyase	Isocitrate = succinate + glyoxalate
5.1.3.1	D-Ribulose 5-phosphate 3-epimerase	Ribulosephosphate 3-epimerase	Ribulose 5-phosphate = xylulose 5-phosphate
6.2.1.2	Acid:CoA ligase (AMP)	Acyl-CoA synthetase	ATP + fatty acid + CoA = AMP + pyrophosphate + acyl-CoA

and bacteria, allows the net synthesis of carbohydrates from fatty acids and is important in the germination of seeds containing reserves of fat. Such tissues also possess an active *lipase* (EC 3.1.1.3) which removes fatty acid residues from the esterified lipid. Two of the other enzymes listed (EC 5.1.3.1 and EC 2.2.1.1) are involved in the *pentose phosphate pathway*, while the last enzyme, *acid:CoA ligase*, is important in fatty-acid biosynthesis.

Before carrying out an enzyme study some attention must be given to the choice of an extraction procedure. With plant materials, particularly leaves and other vegetative tissues, certain difficulties arise in preventing enzyme denaturation or inactivation during extraction. This problem is due chiefly to the formation of oxidation products of endogenous phenols, which inhibit many enzymes and may cause protein precipitation (Anderson 1968). These products include quinones, condensed tannins, and brown pigments, produced mainly as a result of the action of oxidoreductase enzymes such as *o*-diphenol:oxygen oxidoreductase (EC 1.10.3.1). Enzymes of this type, usually referred to as *phenoloxidases*, are brought into contact with their substrates as soon as the normal compartments of the cell are disrupted during maceration or grinding of tissues. The mechanism of enzyme inactivation is probably the reaction of the oxidation products with free sulphydryl groups and terminal and ε-amino groups in polypeptide chains. In order to overcome this problem, protective compounds must often be included in extraction media to inhibit the activity of phenoloxidases. Suitable additives include PVP (polyvinylpyrrolidone), chelating agents such as DIECA (diethyldithiocarbamate), which bind the copper in phenoloxidases, and thiol reagents such as *cysteine*. Thiol compounds are particularly effective in maintaining enzyme activity in leaf extracts.

An alternative approach, which has been used with only limited success, is the removal of phenolic compounds from fresh material with acetone. The resulting *acetone powder* is then extracted with buffer. The major problem is that following the use of acetone many plant enzymes are impossible to dissolve. Furthermore, incomplete removal of phenols may lead to some oxidative browning of the dried powder.

As well as taking steps to prevent the oxidative inactivation of enzyme extracts, care must be exercised to ensure that the extraction medium is of appropriate *pH* and *tonicity*. Enzymes can be extracted into solution from fresh material by manual grinding with a pestle and mortar or by making use of an electrically driven cone-shaped pestle rotating within a stationary glass container. Details of these extraction techniques and comments concerning the choice of buffers were given in section 2.3. In all cases the temperature throughout the extraction procedure must be maintained at around 4 °C. All buffers and glassware should be chilled before use and any subsequent operations should also be carried out at low temperature. Care must be taken to avoid frothing of solutions, for example during the use of electrically driven homogenizers, as proteins tend to precipitate under such treatment. There are certain enzymes which exhibit great stability even in the absence of these precautions, but unless the stability of the enzyme which is being examined is well established, it is certainly advisable to take great care in the preparation and handling of solutions. An additional measure which may be necessary to obtain full enzyme activity is the addition of EDTA (ethylenediamine tetraacetic acid) to overcome inhibition by heavy metal ions.

After disruption of tissue, fibrous material and cell debris are removed by centrifugation and enzyme assays can be carried out on the supernatant. Alternatively, the solution can be fractionated by differential precipitation with ammonium sulphate or by column chromatography. Assays are then carried out on selected fractions. If column chromatography is employed this results in the dilution of enzyme activity and it may be necessary to concentrate samples before analysis. A convenient technique is to place the solution into a length of dialysis tubing which is then placed into a large beaker and covered with solid flakes of polyethylene glycol (molecular weight ~4000). Water is removed from the inside of the tubing and the solution can be concentrated several times within a short space of time. The dialysis should be carried out at low temperature (4 °C).

The activity of enzymes can be measured by following the rate of *disappearance of substrate* or the rate of *accumulation of product*. Many reactions can be studied by means of *spectrophotometry*, either by direct examination of solutions, or by the estimation of absorption at a specific wavelength, following the reaction of an appropriate reagent with the product of the enzyme catalysed process. Measurements of activity should be carried out in reaction mixtures containing the appropriate co-factors, together with an excess of substrate.

The activity of enzymes is expressed in *enzyme units*. One unit is defined as the amount of enzyme which will catalyse the transformation of 1 μmol of substrate per minute under the conditions specified. Plant enzymes are usually assayed at 25 °C. When enzymes are fractionated the degree of purity is usually assessed by measuring the *specific activity*, which is the amount of activity per milligram of total protein.

5.2.5. *Plants as sources of dietary protein*

Plants frequently contribute a major part of the protein intake of man and domestic animals. This is particularly true in areas of the world where primitive farming systems operate at low levels of mechanization and intensification. However, even in Western countries with intensive systems of husbandry, great dependance is placed on plant proteins in the formulation of animal diets. Important sources of protein include pea and bean meals, oilseed residues and fresh, dried, and conserved forage crops. Cereals are also important, as although they usually contain only about 10 per cent protein, they are consumed in large amounts. Because of the fact that many diets are based on cereals, the essential amino acid, lysine, is often present in limiting amounts. Other amino acids, such as methionine and tryptophan, may also be important in certain situations. The total crude protein and lysine contents of some crops and crop products are given in Table 5.4.

TABLE 5.4

Crude protein and lysine content of some crops and crop products

	Protein (% of dry matter)	Lysine (g per 16g nitrogen)
Wheat	10	2·8
Barley	10	3·2
Oats	9	3·4
Maize	10	2·8
Peas†	27	7·7
Beans†	25	7·2
Soya bean meal‡	48	6·7
Groundnut meal‡	50	3·5
Linseed meal‡	34	3·6
Dried grass§	15	7·2

†Whole seeds.
‡Meal remaining after extraction of oil.
§Protein content varies considerably according to stage of growth.

5.3. Other nitrogen-containing compounds

Although most of the nitrogen in plant tissues is usually found incorporated in amino acids and proteins, other important nitrogen-containing compounds are also present. These include the *porphyrins* and the *nucleic acids*. Another group is the *alkaloids*, a heterogeneous family of compounds, many of which exhibit physiological activity when ingested by animals. Some important alkaloids are described in section 5.9.

5.3.1. *Porphyrins*

The molecular structure of the porphyrins is based on the tetrapyrrole ring system with a centrally bound metal atom. Some of the porphrins occur in combination with protein. Examples include the iron containing *haem proteins*, such as the cytochromes. However, in higher plants the most abundant porphyrins are the two ubiquitous pigments, *chlorophyll a* and *chlorphyll b.* The formula of chlorophyll a is shown below. At the centre of the heterocyclic ring

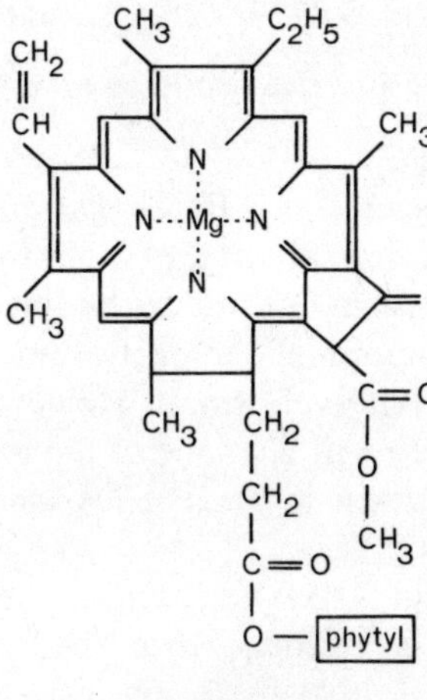

Chlorophyll a

system is a magnesium atom. Protruding from the planar tetrapyrrole ring is a long side-group formed by the esterification of one of the carboxyl groups with the alcohol phytol ($C_{20}H_{39}OH$). It has been suggested that this hydrophobic group is involved in the positioning of the molecule in the chloroplast membranes in a way which ensures the efficient absorption of light by this pigment. The energy 'trapped' by chlorophyll is used in the formation of ATP and NADPH which are utilized in the dark reaction of photosynthesis. This latter process results in the formation of sugars and other organic compounds.

The chlorophylls can be extracted from fresh plant material by homogenization in acetone/water (80:20 v/v). Successive volumes of solvent are used until no traces of green colour are left in the residue. The amount of chlorophyll in the extract can be determined by spectrophotometry. As the two chlorophyll pigments have different absorption spectra, measurements must be made at two wavelengths, 663 nm and 645 nm. The former wavelength corresponds to an absorption maximum for chlorophyll a, the latter to a maximum for chlorophyll b. As there is some overlap in the spectra, a correction factor must be applied to eliminate the interference of chlorophyll b in the estimation of chlorophyll a, and vice versa. The appropriate equations have been worked out by Mackinney (1941).

$$[\text{Chlorophyll a}] = 0{\cdot}0127 \times A_{663} - 0{\cdot}00269 \times A_{645},$$
$$[\text{Cholorphyll b}] = 0{\cdot}0227 \times A_{645} - 0{\cdot}00468 \times A_{663},$$

where A = absorbance and the resulting concentrations are in g/l.

5.3.2. *Nucleic acids*

During the last two decades, it is likely that more has been written about the nucleic acids than about any other topic in biochemistry. Despite the rapid increase in our understanding of the structure and function of these compounds the impact of nucleic-acid research in crop science has been relatively small. However, in future years it is certain that greater use will be made of the wealth of information now available, particularly in such fields as plant breeding and developmental physiology.

Deoxyribonucleic acid (DNA) is found in the nucleus and also in the chloroplasts and the mitochondria. The configuration of the molecule is that of a double helix, the two 'strands' running in opposite directions around a common axis. Each 'strand', or 'chain', consists of a linear polymeric arrangement of alternating sugar (deoxyribose) and phosphate groups with one base attached to each sugar. The bases pair by hydrogen bonding with the appropriate bases in the second strand. Four bases are commonly found in DNA and it is known that *thymine* pairs with *adenine*, and *cytosine* with *guanine*. Thus the amounts of cytosine and guanine found are usually equivalent. However, the amount of (cytosine+guanine) varies as a proportion of the total base content of the DNA. Variations of this type can be used to characterize DNA from different parts of the cell and from different species.

Ribonucleic acid (RNA) is found in the cell as *ribosomal RNA* (rRNA), *transfer RNA* (tRNA), and *messenger RNA* (mRNA). The ribosomal RNA forms a part of the structure of the ribosome whereas the other two types of RNA are concerned with the transfer of materials (tRNA) or information (mRNA) for protein synthesis. The tRNA is concerned with the binding of amino acids prior to peptide bond formation. The individual amino acids are linked together in a sequence determined by the base sequence of mRNA. This process is known as *translation* and the formation of mRNA under the influence of DNA is referred to as *transcription*. It is assumed that the chloroplast contains all three types of RNA as it is able to carry out independent protein synthesis. The bases in RNA are linked by alternating phosphate and *ribose* groups. The four bases present are cytosine, uracil, adenine, and guanine. Other less common bases are also found in some RNAs.

The amount of DNA present in plant tissue varies according to the nature of the cells present. Root tips, pollen, and seeds are cited by Harborne (1973) as rich sources of DNA. The level of RNA fluctuates according to the physiological state of the tissue. During senescence, over three-quarters of the total RNA may be broken down.

Prior to the estimation of the nucleic acids, compounds which may interfere in the analysis are removed with boiling ethanol (Giles and Myers 1966) or ice-cold perchloric acid (Detchou and Possingham 1972). Lipids can be extracted with a mixture of ethanol/ether/chloroform (2:2:1). The solid material is then hydrolysed with trichloroacetic acid (5 g per 100 ml) and the total nucleic-acid content determined by a direct spectrophotometric method involving the measurement of the absorbancy at 268·5 nm (Detchou and Possingham 1972). The DNA content is obtained by measuring colour development following reaction with diphenylamine, and the RNA can then be estimated by difference.

5.4. Lipids

Plant lipids are compounds which can be extracted from fresh or dry material with organic solvents such as chloroform, light petroleum, or diethylether, or with solvent mixtures such as chloroform–methanol.

Lipids serve two major biological roles, namely a structural one as components of membranes, and a storage function, particularly in certain seed tissues. They may also be involved in the enhancement of catalytic activity, in the case of membrane-bound enzymes, and in providing a suitable molecular environment for the operation of energy transfer systems in the mitochondrion and chloroplast.

As sources of oil and fat in animal diets, plant lipids play an important role in nutrition. Many plant lipids contain polyunsaturated fatty acids including linolenic, linoleic, and arachidonic acid. These compounds are essential to normal growth and development in animals. In addition, plant lipids are of high energy value and thus the total lipid content of crop products is of practical significance. The crop scientist usually obtains a routine estimate of total lipid by the gravimetric examination of an extract prepared with diethyl ether or light petroleum. Extracts are prepared in a *Soxhlet* apparatus (Association of Official Agricultural Chemists 1970) which provides for the continual extraction of a dried and milled sample over a period of hours. The apparatus consists of a flask containing boiling solvent and a refluxing system which allows the condensed solvent to flood a glass chamber containing the sample in a cellulose thimble. When the chamber fills to capacity a siphon returns the extract to the flask and the next extraction cycle starts.

5.4.1. *Classes of lipid*

Plant lipids can be grouped, according to molecular structure, into a number of different classes. All compounds within a particular class show certain common structural features. Due to their importance as constituents of commercially exploited plant oils and fats, the *acyl glycerols* are most familiar to many botanists. Many seed tissues contain large amounts of *triacyl glycerols* which contain three fatty acid residues linked to a 3-carbon glycerol backbone.

The general formula of these compounds, frequently known as *triglycerides* is shown below.

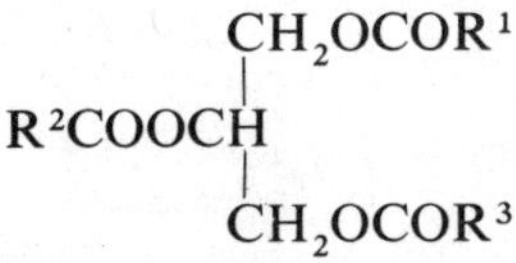

A triglyceride

The R-groups are often polyunsaturated and this results in a lov melting point. Thus many triacyl glycerols are oily liquids at room temperature. *Mono-* and *diacyl glycerols* occur only in very small amounts and are of less importance.

In the plant, triacyl glycerols serve as storage materials which are broken down during seed germination or during other phases of development. In order to utilize seed oils for human consumption, or for manufacturing purposes, the seed tissues must be extracted with a solvent, usually hexane. Alternatively the oil can be expressed by high-pressure and heat treatment. Solvent extraction is, however, the more efficient process and with this technique the residual oil content can be reduced to around 1 per cent. On the laboratory scale, triacyl glycerols can be dissolved in a variety of solvents, including hexane and chloroform. Plant lipids are used in the production of margarine and cooking oils and in a variety of industrial processes.

Glycolipids occur widely in many parts of the plant, usually as components of lipoprotein membranes. They are particularly abundant in photosynthetic tissues, but are also known to occur in such diverse products as potato tubers and apple fruits. Glycolipids contain a carbohydrate moiety linked to the glyceride structure and are frequently identified by colour reactions due to the carbohydrate part of the molecule. The major glycolipids are: *monogalaetosyldiacylglycerol*, *digalactosyldiacylglycerol*, and *sulphonyl-6-deoxyglucosyldiacylglycerol*. The common names and usual abbreviations for these compounds are, respectively: monogalactosyl diglyceride (MGD), digalactosyl diglyceride (DGD), and sulpholipid (SL).

Chloroplasts contain large amounts of galactolipids although the levels vary according to the stage of maturity of the tissue. During

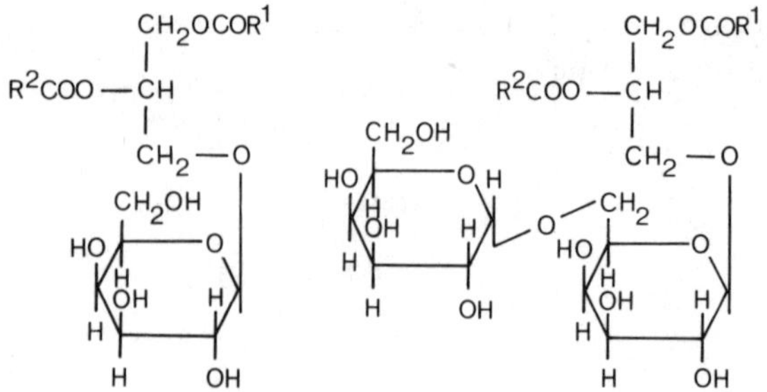

MGD

DGD

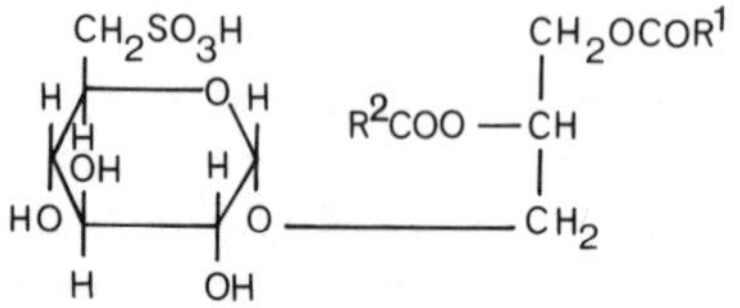

SL

leaf senescence the galactolipids are rapidly broken down with the release of free fatty acids, chiefly linolenic acid.

Another important group of plant lipids are the *phospholipids*. These compounds contain a phosphorus atom and are ubiquitous constituents of membranes. A well-known example is *phosphatidyl*

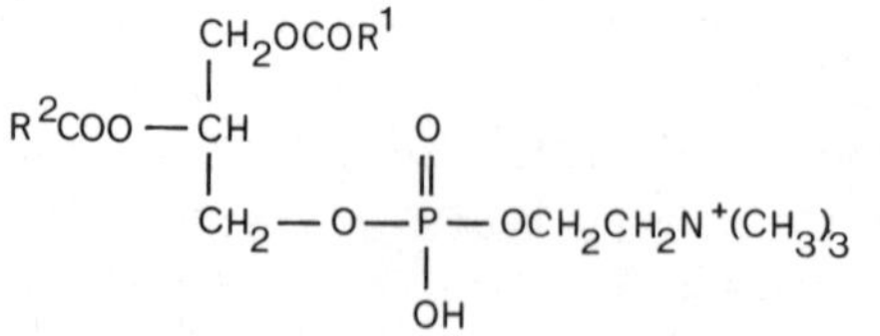

Lecithin

choline, usually referred to by its common name, *lecithin*. Other widely distributed phospholipids include the five compounds illustrated opposite.

Each of the three groups of lipids described above contain fatty-acid residues. A large number of different *fatty acids* are found in plant tissues and each molecular species is characterized by its chain length, and by the number, position, and nature (*cis* or *trans*) of the double bonds. A shorthand notation is often used to describe

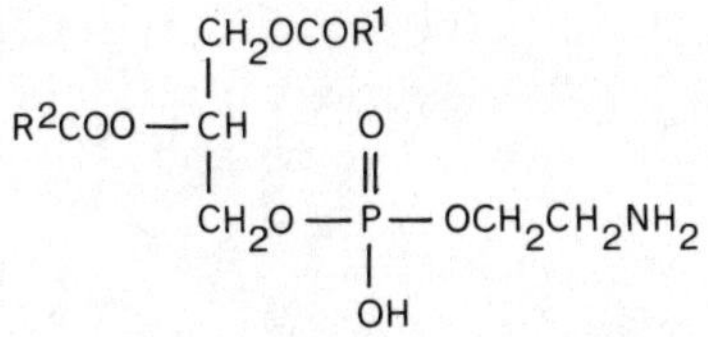

Phosphatidyl ethanolamine

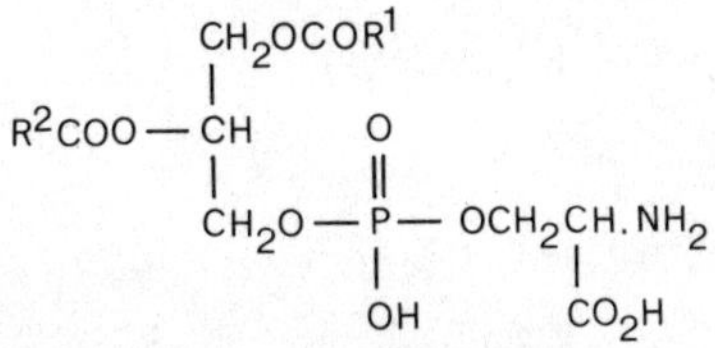

Phosphatidyl serine

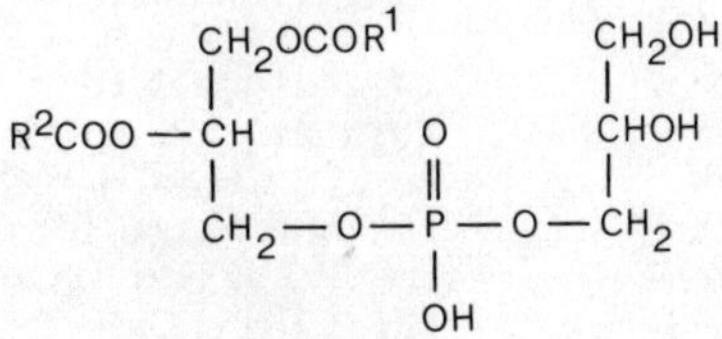

Phosphatidyl glycerol

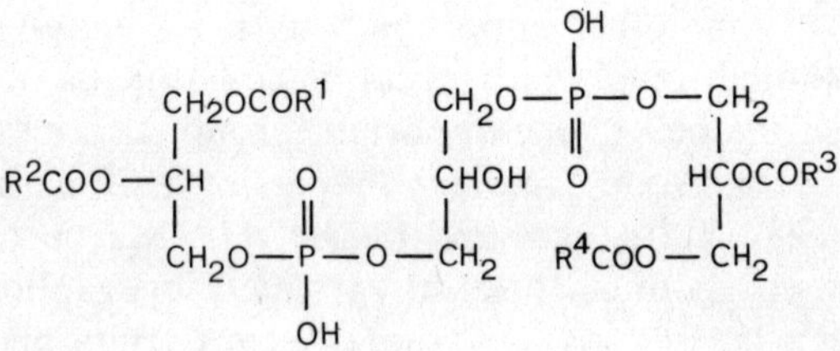

Diphosphatidyl glycerol

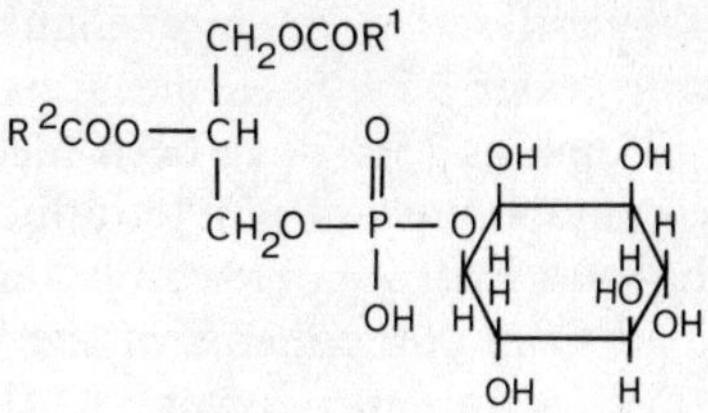

Phosphatidyl-myo-inositol

the structure of fatty acids. Palmitic acid is denoted '16:0', due to the presence of 16 carbon atoms but no double bonds. Linolenic acid, containing 18 carbon atoms and 3 double bonds is thus written '18:3'.

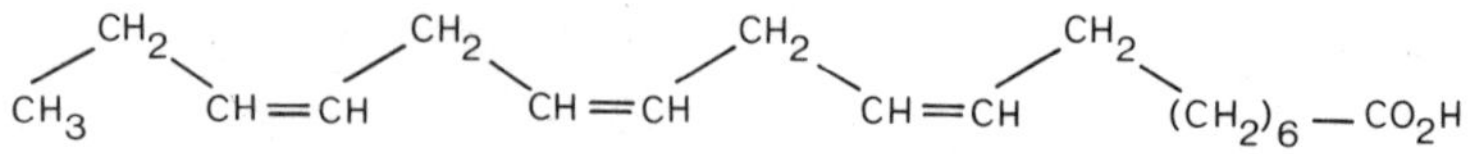

α-Linolenic acid

TABLE 5.5

Formulae of some common fatty acids

Acid	Abbreviation	Formula
Lauric	12:0	$CH_3(CH_2)_{10}CO_2H$
Myristic	14:0	$CH_3(CH_2)_{12}CO_2H$
Palmitic	16:0	$CH_3(CH_2)_{14}CO_2H$
Stearic	18:0	$CH_3(CH_2)_{16}CO_2H$
Oleic	18:1	*cis*-$CH_3(CH_2)_7CH = CH(CH_2)_7CO_2H$
Linoleic	18:2	*cis*-$CH_3(CH_2)_4(CH = CHCH_2)_2(CH_2)_6CO_2H$
α-Linolenic	18:3	*cis*-$CH_3CH_2(CH = CHCH_2)_3(CH_2)_6CO_2H$
Arachidic	20:0	$CH_3(CH_2)_{18}CO_2H$
Arachidonic	20:4	*cis*-$CH_3(CH_2)_4(CH = CHCH_2)_4(CH_2)_2CO_2H$

Most unsaturated fatty acids possess *cis* double bonds. The formulae of some other important fatty acids are given in Table 5.5.

Although most lipids occur as constituents of the cytoplasm or of cell organelles, one important group occurs extracellularly. The *surface waxes* are complex mixtures of hydrocarbons together with long chain fatty acids and their esters (Martin and Juniper 1970). These compounds form an extremely hydrophobic surface which protects the underlying tissues of leaves, fruits, and other plant parts from excessive water loss or uptake and from invasion by pathogenic microorganisms. Often the wax is deposited in a micro-crystalline form and this exaggerates the water repellancy of the epidermis. Contact angles with water have been measured for many leaf surfaces and values as high as 150° have been recorded. These higher values exceed the contact angle of polytetrafluoroethylene (PTFE), which is one of the most hydrophobic compounds known. The contact angles obtained from components of the wax fraction coated on to surfaces artificially are much lower, and thus the hydrophobic nature of the leaf surface is often due as much to the physical pattern of wax deposition as to the molecular nature of the compounds present.

The precise chemical composition of the wax depends on species, environment, and the part of the plant under consideration. The hydrocarbons usually range from 20 to over 30 in carbon number. The major fatty acids are often around 30 carbon atoms in length. Long chain alcohols of similar size are also present as are certain minor components such as the *sterols*, *β-sitasterol*, and *stigmasterol*. These two compounds are also widely distributed in other tissues together with additional sterols.

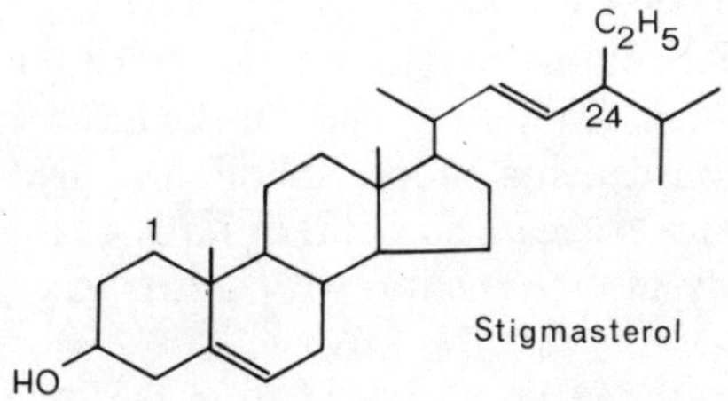

Stigmasterol

5.4.2. *Extraction of lipids*

Procedures for the extraction and analysis of plant lipids have recently been reviewed by Hitchcock and Nichols (1971). These authors emphasize the need for the adoption of appropriate extraction techniques, chosen to avoid alterations to the lipids due to chemical or enzymatic activity. Ideally plant material should be sampled immediately prior to treatment with solvent. Storage at room temperature can result in the breakdown of lipids due to the activity of lipolytic enzymes such as *phospholipase D* and the *galactolipases*. Similar effects are also observed in deep-frozen tissues which are allowed to thaw before extraction.

Leaves and other soft materials can be disrupted by homogenizing the intact tissue with solvent using an electrically driven blender. Top-drive machines are preferable to the normal kitchen models in which the bearing is exposed to solvent. Certain extractants, including chloroform, can exacerbate the problem of lipolysis due to enhancement of enzyme activity. Breakdown of lipids can be avoided by an initial extraction with cold isopropanol (20 volumes) prior to further treatment with a mixture of chloroform and methanol (2:1). This combination of solvents is not only appropriate to leaf tissue but has also been used successfully to remove lipids from cereal grains which must, of course, be milled to a fine meal before treatment.

Following the extraction of a total lipid fraction, it is important to avoid autoxidation of unsaturated fatty-acid residues. Hitchcock and Nichols (1971) recommend storage at low temperature (−5 °C) in solvent containing 0·005% of the antioxidant BHT (butylated hydroxytoluene). This compound can also be incorporated into chromatographic solvents.

5.4.3. *Separation and estimation of fatty acids*

Plant material contains a free-fatty-acid (FFA) fraction and an esterified fatty-acid fraction (EFA). In the latter case the fatty-acid residues are constituents of phospholipids, glycolipids, and acyl glycerols. Surface waxes also contain fatty acids but these will be discussed briefly in conjunction with other wax components, in a later section.

Free fatty acids can be separated from the other major lipids by t.l.c. Glass plates coated with silica gel G to a thickness of at least 0·25 mm, and activated at 110 °C (25 min), are streaked with 0·5–1·0 ml of solution. The plates are developed in a solvent system such as hexane:diethyl ether:acetic acid (80:20:2). The position of FFA is determined using a standard marker run as a spot at each end of the plate. The relevant area of silica gel is scraped from the plate and FFA eluted by shaking with chloroform for one hour, after which the silica gel is removed by paper filtration.

An estimate of total FFA can be obtained by the procedure of Duncombe (1963) which involves the formation of a copper soap followed by extraction into chloroform. The amount of chloroform-soluble copper is estimated spectrophotometrically following the formation of a coloured product with DIECA (sodium diethyldithiocarbamate).

Estimates of individual fatty acids in both the FFA and EFA fractions are usually carried out by g.l.c. Although it is possible to run the free acids on certain columns, it is more usual to methylate the acids in order to increase volatility and improve the separation. The most convenient method for obtaining fatty-acid methyl esters involves the use of boron trifluoride reagent, which consists of 14% BF_3 in methanol and is available from a number of suppliers. The lipid sample is dried down under nitrogen in a small screw-top bottle. A small quantity of reagent is added and after replacing the cap the methylation is carried out at 60 °C (30 min). If rubber lined screw tops are used they should be lined with a non-reactive plastic film such as Melinex polyester (ICI). After cooling, 2–3 volumes of water are added and the methyl esters extracted into light petroleum. In this way methyl esters can be obtained from both the FFA and EFA fractions. Appropriate stationary phases for fatty-acid separations include silicon greases, such as SE 30, and polyesters, such as polyethylene glycol adipate (PEGA). A

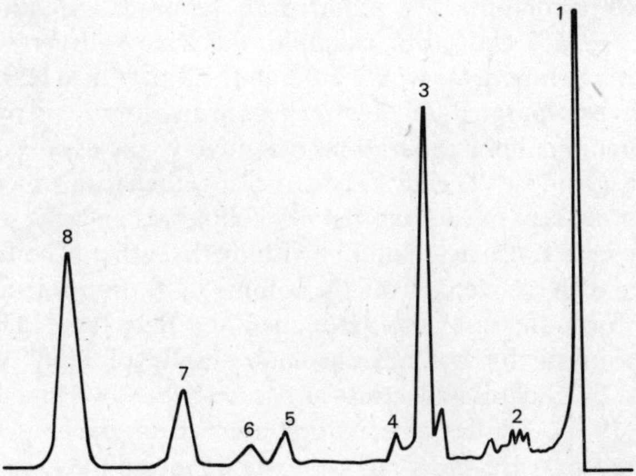

FIG. 5.6. Separation of the fatty-acid methyl esters of palmitic (3), palmitoleic (4), stearic (5), oleic (6), linoleic (7), and linolenic (8) acids. The solvent peak (1) marks the start of the separation. The several small peaks (2) are due to short chain fatty acids.

typical separation from a leaf extract is shown in Fig. 5.6. The stationary phase was 10% PEGA on 100/120 mesh celite and the oven was operated isothermally at 190 °C. The sample injected onto the column was 1·0 μl of a chloroform solution of fatty-acid methyl esters.

5.4.4. *Analysis of acyl lipids*

The major lipids present in chloroform–methanol or isopropanol extracts are usually the phospholipids, glycolipids, and acyl glycerols, together with small amounts of free fatty-acids. Other components also occur, including sterols, sterol esters, and hydrocarbons.

The phospholipids, glycolipids and acyl glycerols form an extremely complex mixture of molecular species. Most analytical schemes are based on t.l.c. but a preliminary column separation may also be required. If this is the case, the separate fractions eluted from the column are then examined individually and a composite picture of the over-all lipid composition of the extract is pieced together. Usually this combination of procedures is sufficient to resolve mixtures to the level of individual classes (e.g. phosphatidylcholine, phosphatidylethanolamine). At this point, fatty-acid analyses can be carried out on specific fractions. However, further specialized techniques are required to separate individual components *within* a class. For example, the triacyl glycerol fraction will contain components which differ in their fatty-acid residues and which can be separated only by further chromatographic treatment.

Preliminary column separations may involve the use of silicic acid or DEAE cellulose. The acyl glycerols, together with other neutral lipids, can be separated from the phospholipids and glycolipids by elution from a silicic acid column with diethyl ether. The latter two classes are then recovered from the column by elution with a mixture of chloroform and methanol (Roughan and Batt 1968). The use of DEAE cellulose for column chromatography of lipids has been developed by Nichols and James and is described by Hitchcock and Nichols (1971). The cellulose is first slurried in glacial acetic acid and poured into the glass column. Methanol and chloroform are then used successively to wash the cellulose prior to the application of a sample of lipids dissolved in chloroform. The different components are eluted with varying proportions of methanol in chloroform. Individual fractions can then be examined by t.l.c. and each resolved class of lipid estimated by a gravimetric or colorimetric method.

Silica gel is used as the thin-layer coating material. A suitable solvent for examining mixtures of glycolipids and phospholipids is chloroform–methanol–acetic acid–water (85:25:8:3 v/v). Similar mixtures can also be utilized as solvents for separating total lipid extracts. Spots can be located by spraying with 25% H_2SO_4 and heating at 200–20 °C. After 10 minutes at this temperature the spots are completely charred and the plates can be evaluated on a semi-quantitative basis by densitometry. A typical separation of leaf lipids, recorded as a densitometric trace, is shown in Fig. 5.7. During the heat treatment of the sprayed plates, characteristic colours are formed. Phospholipids and acyl glycerols give brown

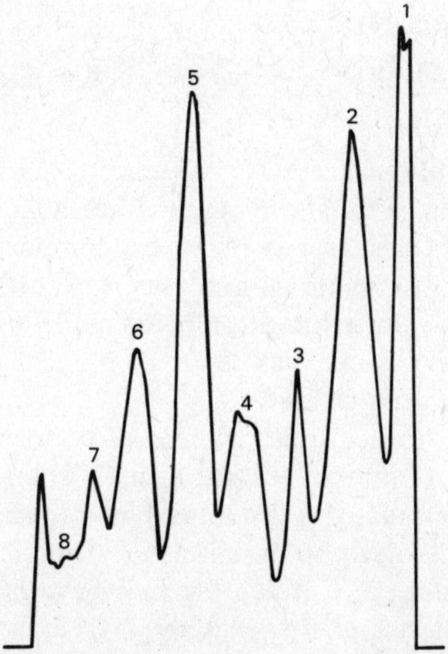

FIG. 5.7. Densitometer trace of lipids extracted from leaf tissue and separated by t.l.c. on silica gel. The solvent flow was from left to right. The positions of the various lipids are as follows: (1) pigments and neutral lipid, (2) MGD, (3) sterol glycoside, (4) phosphatidylethanolamine, (5) DGD and phosphatidylglycerol, (6) phosphatidylcholine (7) sulpholipid, (8) phosphatidylinositol.

spots and the galactolipids a purple coloration. Alternative methods of locating the separated components include the use of iodine vapour or the spray reagent, dichlorofluorescein.

For accurate quantitative analyses of the galactolipids a preliminary column separation on DEAE cellulose is necessary. This is followed by t.l.c. and estimation of the galactose by the procedure of Roughan and Batt (1968). Phospholipids can be determined by the method of Rouser, Siakatos, and Fleischer (1966) following similar fractionation procedures. For the separation of lipids according to the nature of their fatty-acid residues, plates are coated with silica gel containing silver nitrate. Separation within lipid classes then depends on the number and position of the double bands present (Hitchcock and Nichols 1971).

In addition to the use of t.l.c. and low performance columns, it is also possible to separate lipid classes by high pressure liquid chromatography (h.p.l.c.). Commercially built units are now available but their relatively high price is a disadvantage compared with more conventional methods. However, the convenience of

BANGOR LIBRARY

avoiding the transfer of sample components from developed thin layer plates into solution once again, offers a very real advantage.

5.4.5. *Surface waxes*

The examination of surface waxes requires special extraction techniques chosen to ensure that only the extracellular surface lipids are recovered. Much of the early work on wax components, particularly fatty acids, is of little value as analyses were often performed on total tissue extracts, the assumption being made that leaves or other tissues with a heavy outer wax layer contained only insignificant amounts of lipid within the cells.

To remove the surface lipids, whole leaves or fruits are simply dipped in several volumes of solvent, usually chloroform, for about two minutes. The extract is then reduced in volume and fractionated by column chromatography on silica gel. Detailed separations of the wax of a number of different species have been described recently (see Hamilton and Power 1969; Macey and Barber 1970). After a preliminary separation into lipid classes, g.l.c. is used to identify individual hydrocarbons, alcohols, esters, and fatty acids. For some of these compounds, high operating temperatures (200–300 °C) are required due to their long chain length.

5.5. Sugars and storage polysaccharides

As immediate products of photosynthesis, sugars and storage polysaccharides occupy an important position in the energy economy of the green plant. Free sugars are translocated from the shoot system to heterotrophic regions of the plant, where they supply the organic building blocks required in biosynthesis, while storage carbohydrates represent a means of conserving energy, as the simple sugars released upon their hydrolysis can be utilized for cellular respiration.

To the crop scientist, sugars and storage polysaccharides are important for two main reasons.

(a) The presence of sugars in certain amounts and proportions may be related to distinct physiological events and chemical analysis may thus lead to further evaluation or understanding of the processes concerned. For example, cold-hardiness in some species is associated with high levels of sugars and thus sugar content can be used as an indicator of the progress of cold-hardening.

(b) The economic value of some crops and crop products may depend, in part, on their sugar or starch content. This comment applies both to products such as malting barley, which are used in the food and drink industries, and to crops intended for animal feeding or other purposes.

5.5.1. *Occurrence and distribution*

The most important sugars found in an uncombined form are the *monosaccharides, glucose* and *fructose*, and the *disaccharide, sucrose* (Fig. 5.8). Other compounds occur in smaller amounts including *xylose*, *rhamnose*, and *galactose* (Fig. 5.9), together with a number of sugar phosphates.

Storage polysaccharides are formed in the plant by the linkage of simple sugars to give polymeric structures. The most important storage compound is *starch*. In photosynthetic tissues it is laid down as granules in the chloroplasts. However, it also occurs in non-photosynthetic organs such as tubers, fruits, and rhizomes, often in massive amounts. Starch, in fact, consists of two polysaccharides. *Amylose* molecules are constructed of glucose units joined by

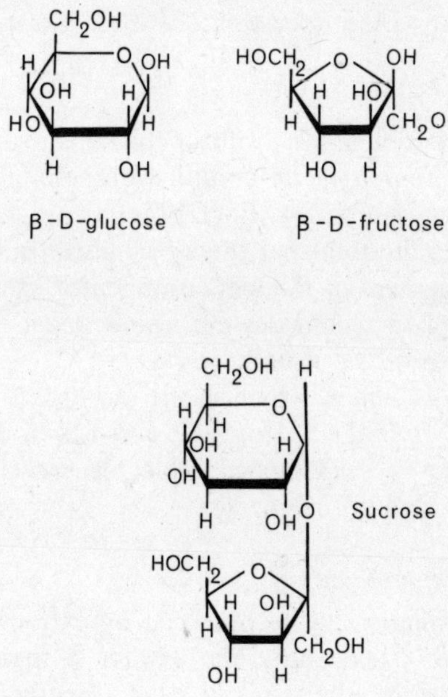

FIG. 5.8. The major sugars occurring as free molecular species.

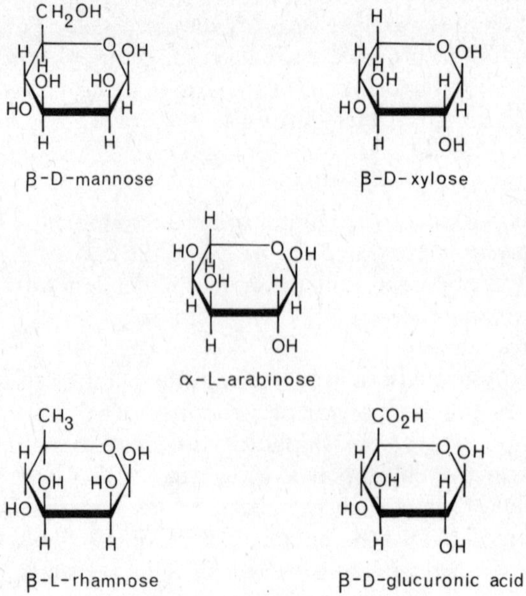

FIG. 5.9. Other monosaccharides found in plant tissues.

α1 → 4 linkages to give long unbranched chains of up to 1000 units in length, while *amylopectin* contains branching points due to the presence of α1 → 6 linkages. Each branch contains between 18 and 27 residues. It is thought that these branched molecules are formed by the rearrangement of the less complicated structure of amylose. Other storage polysaccharides are also known. *Fructans* are made up of fructose residues usually joined by β2 → 1 links. The best-known fructan is *inulin*, which is the storage compound found in the Jerusalem artichoke rhizome. It consists of 30–35 units linked together to give an unbranched molecule. Fructans also occur as storage compounds in the *grasses*.

5.5.2. *Extraction and analysis*

A solution of sugars can be obtained by extracting fresh or dried material with 95% ethanol. The extract is then concentrated by rotary evaporation, filtered, and used directly for qualitative or quantitative analysis. Separation of sugars can be accomplished by chromatography on either paper or an appropriate thin layer. Gas–liquid chromatography is also available as an alternative technique.

The paper chromatography of plant sugars has recently been described by Harborne (1973), who recommends the use of Whatman No. 1 paper developed in a descending fashion and employing a single dimension. Standards are prepared in 10% isopropanol and the chromatogram is developed for up to 36 hours. The paper is then dried, dipped in a solution of aniline hydrogen phthalate, and after evaporation of solvent, is heated at 105 °C. The various sugars give characteristic colours when treated in this way. Quantification can be achieved by eluting the coloured spots with stannous chloride solution and then measuring absorbance at the appropriate wavelength (Harborne 1973).

Thin layer chromatography can be carried out on cellulose or silica gel and separation can often be achieved within a few hours. Gas–liquid chromatography involves the preparation of derivatives, such as the trimethylsilyl ethers. Separations are usually carried out on a silicone polymer stationary phase and flame ionization is used as a means of detection (Holligan 1971).

Estimates of the total sugar content of extracts can be carried out by the anthrone method (Ministry of Agriculture, Fisheries, and Food 1973). This procedure is also used to measure total water-soluble carbohydrate in leaves. This fraction is of significance in nutrition and it may also give an indication of the physiological condition of the plant in relation to cold-stress. Dried or fresh samples are extracted with water for a period of up to 1 hour. Fresh material should be extracted at high temperature to ensure that denaturation of enzymes occurs, thus avoiding changes during extraction. Samples of the extract are heated with the anthrone reagent in a boiling water bath and the absorbance at 620 nm is determined with a spectrophotometer. The system is calibrated with standard solutions of glucose.

Polysaccharides can be determined by assay of the sugars liberated upon hydrolysis. The determination of starch has been reviewed by Whelan (1955) who describes an extraction procedure based on dissolution in perchloric acid. Following purification, the starch is hydrolysed with hydrochloric acid and the glucose estimated by means of the Somogyi method (Somogyi 1945). An alternative procedure, involving the enzymatic breakdown of starch, has been described by MacRae and Armstrong (1968).

5.6. Cell wall materials

External to the cell membrane of most plant cells is a more rigid structure known as the *cell wall.* In stem tissues and in the leaves of grasses and other plants the cell walls often become heavily reinforced to provide the structural support and rigidity necessary to maintain an erect habit of growth. A range of important polymeric materials are found in the cell wall. They include *cellulose*, a number of *non-cellulosic polysaccharides*, and the phenylpropanoid polymer, *lignin.* In addition, a certain amount of protein is present, part of which consists of enzyme-protein bound to the cell wall, while the remainder appears to fulfil some structural role. The latter fraction is rich in 4-*trans*-hydroxy-L-proline. This amino acid is bound to L-arabinose in the cell wall polysaccharides. The resulting *glycoprotein* complex is sometimes known as *extensin* due to its supposed importance in cell extension growth.

5.6.1. *Cell wall polysaccharides*

The major polysaccharide in cell walls of higher plants is *cellulose.* This compound consists of an unbranched arrangement of glucopyranose residues linked by β1 → 4 bonds. The repeating unit thus consists of the *cellobiose* residue shown below.

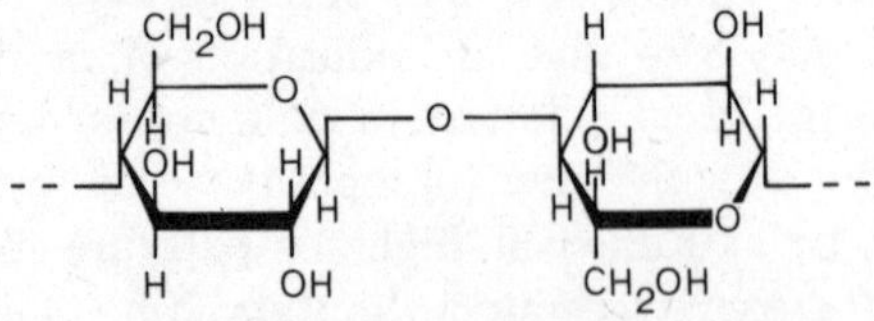

The cellobiose residue

The aggregation of cellulose molecules in an ordered fashion gives rise to microfibrils within the wall. These microfibrils are surrounded by a matrix of *non-cellulosic polysaccharides* consisting of a number of different polymers grouped together as *hemicelluloses.* The latter compounds are soluble in alkaline solutions and this property is one of their few common features. The constituent monosaccharide residues occurring in hemicelluloses include mannose, xylose, glucose, galactose, and arabinose.

A third group of polysaccharides is also found in plant cell walls mainly as an intercellular 'cement' holding adjacent cells together.

This part of the cell wall is known as the *middle lamella* and the polysaccharides are grouped together as 'pectin' or '*pectic substances*'. A number of different components occur but the major one is a straight chain polymer containing D-galacturonic acid residues, linked by α1 → 4 bonds, and known as *galacturonan*, or polyuronide. The basic repeating unit is shown below.

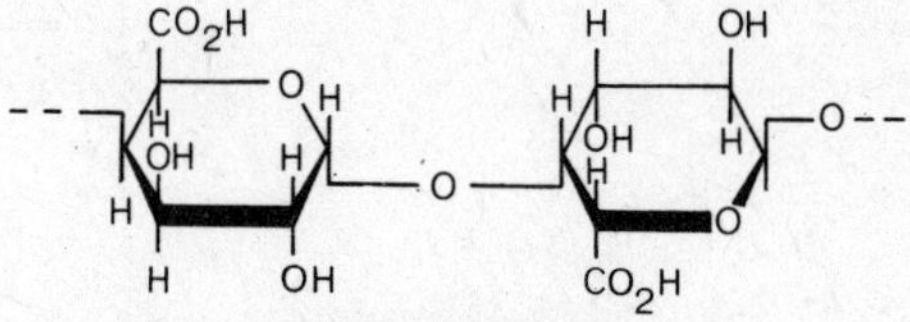

The repeating unit of galacturonan

In fact, the structure is slightly more complex than this formula suggests, as some of the acid groups are present as methyl esters. A summary of the characteristics of the major cell wall polysaccharides is given in Table 5.6.

TABLE 5.6
Cell wall polysaccharides

Polysaccharides	Monosaccharide residue(s)	Nature and function
Cellulose	Glucose	Microfibrillar structure confers important structural properties on the cell wall
Xylans	Xylose	Form the matrix surrounding the cellulose microfibrils. Often known as 'hemicelluloses'
Mannans	Mannose with a small proportion of glucose	
Glucomannans	Glucose and mannose	
Galactans	Galactose with a small proportion of arabinose	
Arabogalactans	Arabinose and galactose	
Galacturonan	Galacturonic acid	The major pectic substance present in the middle lamella

5.6.2. *Lignin and fibre*

Lignin is an amorphous material which encrusts the cellulose fibrils present in the cell walls of certain tissues, notably the vascular elements of stems and leaves. It occurs in particularly large amounts in woody plants and in grass crops which have been harvested too late in the growing season. The term 'lignin' is used to describe a range of similar polymeric materials which are insoluble in 72% sulphuric acid and all contain phenylpropanoid residues as the basic building blocks. These monomer units include compounds such as sinapyl alcohol.

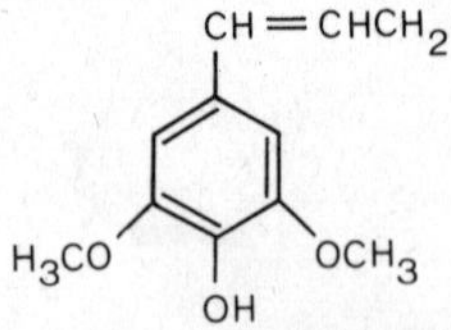

Sinapyl alcohol

Lignin is an extremely inert substance which is highly resistant to the action of enzymes, including those present in the digestive systems of farm animals. Its presence thus decreases the *digestibility* of feedstuffs. The estimation of lignin usually involves the removal of other components with a series of reagents, a process which is time consuming and laborious. This is unfortunate as the routine estimation of lignin would otherwise provide a convenient means of predicting the digestibility of herbage. A recently published technique involving the extraction of lignin with acetyl bromide offers the possiblity of more rapid analysis (Morrison 1972). However, an alternative approach is usually adopted. When dried plant material is refluxed with a strong reagent such as trichloroacetic acid, much of the dry matter is dissolved leaving an insoluble residue of ash and '*fibre*'. Other procedures for fibre estimation involve the use of strong acid followed by strong alkali ('*crude*' fibre), or hot detergent mixed with a mineral acid. The organic matter in the residue consists of a mixture of cellulose, hemicellulose, and lignin. The amount of fibre can thus be determined fairly easily and as it correlates negatively with digestibility it is a useful routine predictor of the nutritive value of leafy crops. Although such analyses are well suited to the rapid evaluation of herbage, it is sometimes the case that an accurate estimation of lignin is required, in which case the more elaborate procedures must be followed.

5.6.3. *Analysis of cell wall materials*

A number of methods are available for the determination of cellulose. A straightforward procedure, described by Crampton and

Maynard (1938), involves the extraction of dried material with a boiling mixture of acetic and nitric acids. The residue is washed with solvent and dried. The organic material, which is taken as cellulose, is determined by the loss of weight on ignition.

Hemicellulose can be determined by extracting a preparation of 'holocellulose' with a solution of sodium hydroxide (4g per 100 ml). The hemicellulose is precipitated from the filtrate with ethanol (Weihe and Phillips 1947). The original preparation of holocellulose is obtained by the extraction of dried material successively with solvent, ammonium oxalate solution, and acid pepsin (Waite and Gorrod 1959).

Pectin can be isolated from suitable tissues by extraction into boiling water. The solution is then neutralized with ammonium hydroxide, concentrated *in vacuo*, and the pectic substances precipitated with alcohol (Harborne 1973).

Many different procedures have been described for the estimation of lignin. One standard procedure depends on the removal of non-lignin components with a series of reagents. The sequence is as follows:

(a) reflux in a Soxhlet apparatus with a mixture of ethanol and toluene,
(b) wash with ethanol and ether,
(c) digest with acid pepsin to remove protein,
(d) wash residue with hot water,
(e) reflux with dilute sulphuric acid,
(f) wash with ethanol and ether,
(g) digest with strong sulphuric acid,
(h) wash with hot water,
(i) reflux with dilute sulphuric acid,
(j) wash with hot water.

The full details are described by Ellis, Matrone, and Maynard (1946) and Waite, Johnston, and Armstrong (1964).

A procedure for the *detection* of lignin in milligram quantities of plant material has recently been developed in the author's laboratory. The technique employed is pyrolysis gas chromatography (p.g.c.). A small sample is pyrolysed and the volatile products are swept onto the chromatograph in a stream of carrier gas. The pattern of peaks traced out on the chart recorder can be used as a 'fingerprint' of the sample (Fig. 5.10). Certain of the peaks are

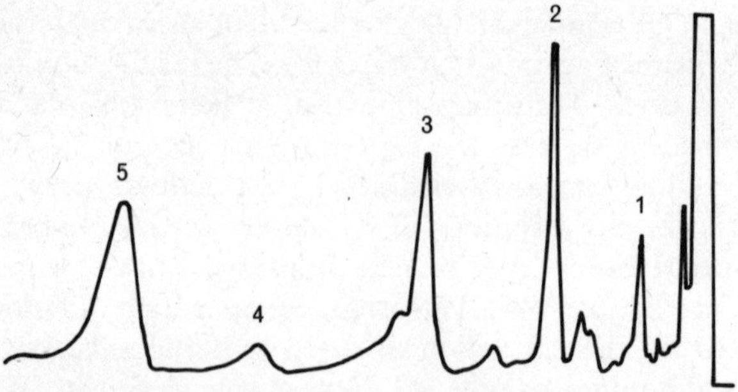

FIG. 5.10. Gas–liquid chromatography of the pyrolysis products of lignin. A small sample was pyrolysed and the volatile products swept onto the column by a stream of carrier gas. The liquid phase was Apiezon L and the column was operated at 135 °C. The components were detected at the column outlet by flame ionization. Several of the numbered peaks, especially (3) and (5), are also observed when whole plant material containing lignin is pyrolysed in a similar manner.

characteristic of lignin. Unfortunately, the technique is difficult to operate in a quantitative manner. However, the method has been used to identify lignin in different samples of soil organic matter.

For the determination of fibre, a recent method is to extract the dried sample with dilute acid and detergent to give a residue which is washed with water and acetone prior to drying. The *acid-detergent fibre* is then determined gravimetrically (Clancy and Wilson 1966). For a discussion of the applications of these procedures in animal nutrition, reference should be made to reviews by Harkin (1973) and Barnes (1973).

5.7. Plant hormones

Plant hormones are endogenous compounds capable of regulating certain aspects of growth or development when present at very low concentrations. Although they form a rather diverse group in terms of molecular structure, their treatment under a common heading is justified by the fact that, in function, they often interact, or at least influence similar or related events. For this reason the botanist must often measure the level of more than one hormone when trying to understand some particular phase of the life-cycle.

Evidence supporting a particular role for a plant hormone is usually derived either from studies of the effect of an exogenous supply of the compound or a close analogue, or from an examination of endogenous levels at different stages of development. Sometimes the evidence is conflicting. For example, the application of synthetic auxins to fruits often promotes growth, but there is little correlation between the natural level of auxin and fruit size. The implication is that other hormones are also involved in fruit development. Similar conclusions relating to other processes are widely reported in the literature and thus great difficulty exists in interpreting patterns of crop growth in terms of hormone activity. However, the increasing importance of synthetic growth regulators and the rapid development of plant breeding programmes which, knowingly or unknowingly, aim at modifying the hormonal status of varieties, point to a need for greater understanding of the natural processes influencing whole-plant development.

5.7.1. *Molecular structure and biological properties*

The term 'plant hormone' is correctly used only in relation to endogenous compounds. Many synthetic hormone analogues are known, and frequently these compounds have been widely studied, simply due to their ready availability. Examples include the herbicide 2,4-dichlorophenoxyacetic acid (2,4-D) and the purine kinetin. These compounds possess activities similar to those of the naturally occurring auxins and cytokinins respectively. Important examples from these groups are shown below (1 and 2).

1. Auxins

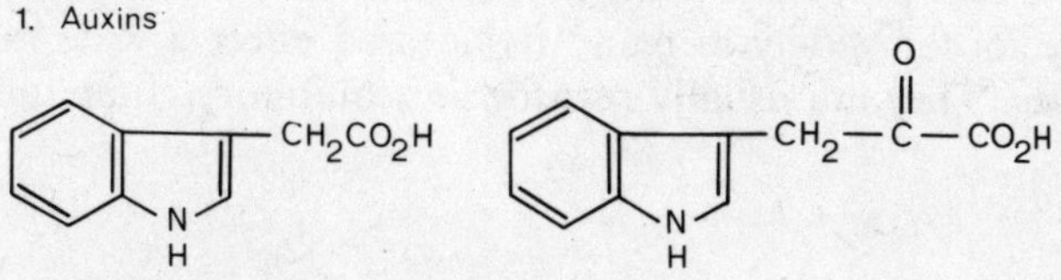

Indoleacetic acid Indole pyruvic acid

2. Cytokinins

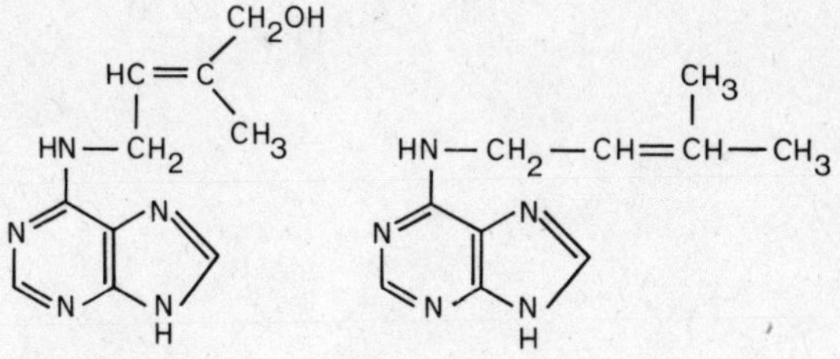

Zeatin 6-(γ,γ-dimethylallylamino) purine

A third group of hormones, *the gibberellins*, consists of a large number of compounds whose physiological activities and chemical identities are well established. The molecular structure of these compounds is based on the *gibbane skeleton*.

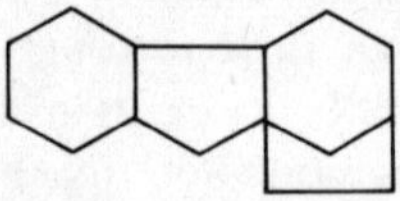

The gibbane skeleton

The first gibberellin to be isolated and characterized was *gibberellic acid* (GA_3) which is widely distributed in plant tissues.

3. Gibberellins

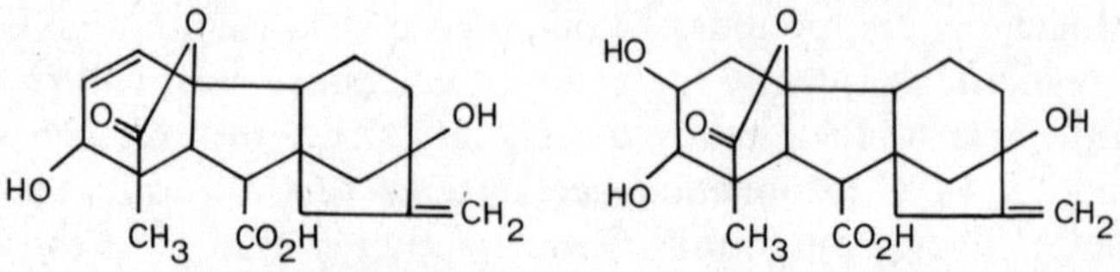

GA_3 GA_8

Two other important plant hormones, namely *abscisic acid* and *ethylene*, occur widely in many tissues and affect a wide range of processes. They are usually regarded as inhibitory compounds.

4. Abscisic acid

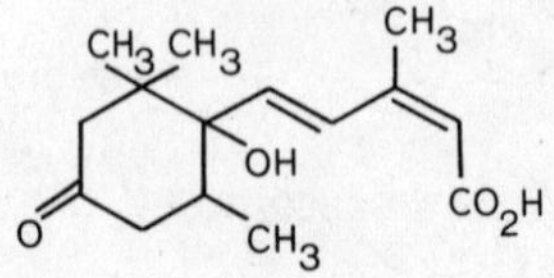

5. Ethylene

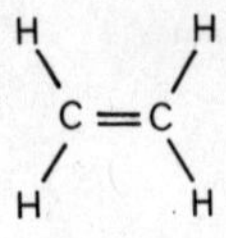

Although the mode of action of many hormones is not clearly understood their effects on whole plants and on plant tissues are extensively documented. The *auxins* are a group of compounds which cause a variety of growth effects frequently involving the stimulation of *cell extension*. They are intimately involved in morphogenesis and in the control of developmental processes such as leaf abscission. The principle auxin in most plants is *indoleacetic acid* (IAA). Compounds related to this parent molecule have been reported but some of these are undoubtedly artefacts produced during the extraction procedure. However, a range of authenticated auxins other than IAA have been characterized. The activity of these compounds, which include *indole acetaldehyde* and *indole pyruvic acid*, is often due to their conversion to IAA within the plant cells.

TABLE 5.7
Processes affected by plant hormones

Auxins	*Cytokinins*
Cell extension	Cell division
Differentiation	Leaf senescence
Tropic responses	Differentiation
Apical dominance	Apical dominance
Root growth	Tropic responses
Fruit development	
Abscission	
Gibberellins	*Ethylene*
Activity of sub-apical meristem	Fruit ripening
Cell extension	Abscission
Expression of genetic dwarfism	Flowering
Tropic responses	Tropic responses
Fruit growth	
Bud dormancy	*Abscisic acid*
Seed germination	Dormancy
Apical dominance	Senescence
Flowering	Rate of growth

The *cytokinins* are involved in the control of *cell division* and high levels are often found in tissues in which this process is occurring rapidly. Examples include germinating seeds and fruits in their initial stages of development. Cytokinins also delay the onset of senescence in leaves and can influence the movement of organic compounds.

Gibberellins occur widely in many tissues and are found in particularly high levels in immature seeds. Over 30 separate gibberellins have been characterized and up to 13 different examples have been found in a single species. The gibberellins are probably best known for their dramatic effect when applied to genetically dwarfed plants of various species. The result is internode elongation and a return to full height. Many other effects have been recorded and some are listed in Table 5.7, together with some activities of the other major hormones.

Abscisic acid (ABA) is a growth inhibitor which has been isolated from the cotton plant, in which it promotes abscission, and from a variety of other sources including fruits, leaves and rhizomes. It is involved in the maintenance of bud dormancy in certain woody plants and in a variety of other phenomena including cold-hardiness, wilting, and stomatal closure. Another important inhibitor is the hydrocarbon gas, *ethylene*. The activity of this compound is remarkable in view of its simple molecular structure. Among the processes affected are fruit-ripening, abscission, root growth and, in certain cases, flowering.

5.7.2. *Extraction and sampling*

Prior to assay, an extract of the hormone must be prepared and, if appropriate, some form of purification procedure may be employed. In a recent text, Weaver (1972) has described extraction techniques for the major groups of endogenous growth compounds. Auxins can be extracted in organic solvents such as chloroform, ethanol, methanol, or diethyl ether. Some of the auxin is often present in a 'bound' form and is extracted only after prolonged treatment of tissue. The role of bound auxin is not yet entirely clear and thus results obtained in a subsequent bioassay should always be interpreted with reference to the extraction procedure employed. Fractionation of auxin extracts can be carried out by paper chromatography. The paper is then cut into strips, each one is eluted, and the extracts used for bioassay. Thin layer and column chromatography have also been employed and they may be used as complementary or alternative techniques to paper separations.

Gibberellins can be extracted with mixtures of acetone and water, ethanol and water, or methanol. Procedures for purification usually involve removal of solvent followed by acidification of the aqueous phase and extraction into ethyl acetate, followed by chromatography (West and Reilly 1961).

Cytokinins can be extracted with aqueous or absolute ethanol and fractionated by paper chromatography using a mixture of sec-butanol and NH_4OH (4:1), as described by Van Staden (1973).

Abscisic acid has been extracted in a number of different solvents and like the other acidic hormones it can be purified by solvent partition and chromatography. Sampling procedures for the other important inhibitor, ethylene, may consist simply of the withdrawal of a gas sample using a syringe or alternatively they may involve a method of concentrating the hormone. Hulme and co-workers employ a trap system in which the ethylene is condensed at liquid oxygen temperature and then vaporized directly into the chromatograph using rapid ohmic heating (see for example, Galliard, Rhodes, Wooltorton, and Hulme 1968).

5.7.3. *Bioassay*

Because of their extremely low concentration in most tissues, many hormones are commonly assayed by measurement of their activity in a closely controlled biological system. Often the method used is one of a number of standard procedures but when some new aspect of hormone activity is being studied it may be necessary to select a novel system.

One of the classical effects of *auxins* is the stimulation of cell extension growth. This property is measured in a number of different bioassays including those in which curvature is induced in oat coleoptiles or split segments of pea stem. These methods necessitate the rather problematic measurement of the degree of curvature. The *Avena coleoptile straight growth test* avoids this difficulty. Seeds of *Avena*, for preference a naked type (*A. nuda*), are allowed to germinate for about 72 hours, with exposure to red light. The coleoptile containing the first developing leaf is then removed from the seed and the apical 3 mm is cut off and discarded. Sections exactly 4 mm in length are then taken for the test (Weaver 1972). Removal of the apical tip prevents stimulation of growth by endogenous auxin which is produced in this region. The sections are floated on a solution containing the extracted hormone and the extension growth after 24 hours is compared with that induced by solutions containing known amounts of the reference auxin, indoleacetic acid. This technique can also be used as a general screening test for inhibitors.

Solutions of the *cytokinins* are assayed by an assessment of either their senescence-retarding activity or their ability to promote cell division in callus tissue. For measurement of the *delay of senescence*, leaves of radish or cocklebur are used. Either whole leaves or leaf-discs are treated with the test solution and stored for two or three days in a humid atmosphere. The extent of leaf senescence is measured by visual appraisal or by estimating the loss of chlorophyll. In each case the effect of the solution under assay is compared with the effects of standard solutions of the synthetic cytokinin, kinetin (Osborne and McCalla 1961). The performance of *callus growth tests* requires facilities for the sterile culture of callus cells, using either tobacco stem pith or soya bean cotyledon tissue as a starting material. In the absence of either a cytokinin, or a synthetic analogue in the culture medium, the growth of the callus is prevented. Details of these procedures, developed originally by Skoog, Miller, and others, are given by Weaver (1972). The concentration of cytokinin is assessed by comparison of test solutions with standards consisting of solutions of kinetin within an appropriate concentration range.

Gibberellins can be estimated by measuring the effect of extracts on *genetically dwarfed plants*. For example, the ability of samples to promote leaf sheath growth in dwarf maize seedlings can be related to the gibberellin content of the sample. In this test a micropipette is used to apply 0·1 ml of solution to the top of the shoot. Uptake is facilitated by the inclusion of the detergent Tween 20 in each sample to be analysed. An alternative test is the *barley endosperm assay*. This involves the measurement of amylase-induced sugar production in barley seeds. Amylase activity is promoted by gibberellins and thus the rate of hydrolysis of starch is related to hormone concentration. In order to carry out the test barley grains are cut in half transversely. The section containing the embryo is discarded and the other half is used in the assay. The pieces of endosperm are incubated for 24 hours and the sugar released is estimated colorimetrically.

Problems are often encountered in the estimation of *inhibitory compounds* using bioassay techniques due to the difficulty of finding a sufficiently specific test. *Abscisic acid* was originally detected and isolated by several independent groups of workers and the bioassay technique varied according to the plant species used as a source of the hormone. Wareing and his associates monitored the dormancy inducing activity of abscisic acid in buds of sycamore (Eagles and Wareing 1964). Other tests which have been used are based on such phenomena as the acceleration of abscission or the inhibition of cell extension growth. *Ethylene* produces a range of effects in different tissues but is most easily estimated by direct physico-chemical assay.

5.7.4. *Chemical and physico-chemical analysis*

Although almost all the early work with plant hormones was carried out using bioassay techniques, more recent studies have made use of a range of chemical procedures. *Gas–liquid chromatography* has been applied successfully to the analysis of gibberellins following the preparation of volatile ester derivatives (Cavell, Macmillan, Pryce, and Sheppard 1967). Restricted use of this technique has been made with the cytokinins (Upper, Helgeson, Kemp, and Schmidt 1970) and the inhibitor, ethylene, is routinely determined in this way using a flame ionization detector and a column operated at room temperature (Galliard *et al.* 1968). *Abscisic acid* can be determined by g.l.c. after conversion to its trimethylsilyl derivative (Davis, Heinz, and Addicott 1968). This hormone can also be examined by a technique which makes use of its optical activity (Millborrow 1967).

Colorimetric and fluorimetric procedures have been applied chiefly to the indole hormones. The colorimetric assay of auxin usually depends on either the Ehrlich or Salkowski reaction. A discussion of these methods has been presented by Weaver (1972).

5.8. Inorganic constituents

The inorganic constituents of crops and their products vary in amount according to the stage of maturity of the crop, its conditions of growth, and the nature of the soil. Analyses of mineral elements can serve as a useful indicator of deficiency situations during the growing period. They are also useful in assessing the nutritional status of the harvested crop in relation to the feeding of farm animals.

5.8.1. *Elements essential for plant growth*

In addition to carbon, oxygen, and hydrogen, there are six *major elements* required for the healthy growth of plants, namely: nitrogen, phosphorus, potassium, sulphur, magnesium, and calcium. There are also a number of elements which are required in much smaller amounts. These are the *trace elements*, or micronutrients. They include iron, manganese, boron, zinc, copper, molybdenum, and chlorine. Some plants also show a requirement for sodium and others are possibly dependent on selenium for optimum growth. Cobalt is required for the active symbiosis of legume species with their appropriate root-nodule bacteria. This is due to the fact that vitamin B_{12} is involved in the nitrogen-fixing process.

TABLE 5.8

Biochemical basis for the essentiality of the major elements in higher plants

Element	Role
Nitrogen	Constituent of many organic molecules including proteins, nucleic acids, alkaloids, and porphyrins.
Phosphorus	Present in nucleic acids, nucleotides, some coenzymes, and phospholipids.
Potassium	Important as an enzyme activator.
Sulphur	Constituent of the amino acids methionine and cysteine and also of certain coenzymes.
Magnesium	Occurs in the chlorophyll molecule and also acts as an activator of many enzymes, especially those connected with carbohydrate metabolishm.
Calcium	Present as a constituent of the cell wall. It also functions as an activator of a number of enzymes, including lipoxidase and several phospholipases.

TABLE 5.9

Biochemical basis for the essentiality of the trace elements in higher plants

Element	Role
Iron	Present in a range of enzymes and electron carriers.
Manganese	Activator of many enzymes including some of those required for the Krebs cycle and for glycolysis.
Zinc	A constituent of a number of metalloproteins.
Copper	Required by some enzymes associated with oxidative conversions utilizing molecular oxygen.
Molybdenum	A constituent of nitrate reductase.
Boron	The function of this element is not fully known. It appears to serve a role in sugar metabolism.
Chlorine	Required for the operation of photosystem II in photosynthesis.

The essentiality of the major and trace elements is due either to their importance as constituents of complex molecules or to their role as enzyme activators. The various functions of each element are summarized in Tables 5.8 and 5.9.

5.8.2. *Analysis*

The estimation of metal elements present in plant material is carried out on the residue which remains after a suitable preliminary treatment to remove organic matter. Two alternative procedures are available for the destruction of organic components. The simplest approach is *dry combustion*. Samples of fresh material are dried and an accurately weighed aliquot is heated overnight at 450 °C, in a muffle furnace. The greyish residue is then treated with hydrochloric acid to dissolve the mineral constituents. An alternative method is the *wet-digestion* technique which involves the use of a mixture of nitric and perchloric acids. This procedure can involve the risk of explosion and it is not suitable for materials containing more than 10 g per 100 g dry matter of fat. After digestion the mixture of acids is removed by boiling and the residue is treated with hydrochloric acid to obtain a solution.

Methods available for the quantitative assay of mineral elements include colorimetry, spectrophotometry, and atomic emission or absorption analysis (section 4.1). In order to assess the applicability of these methods to individual elements, reference should be made to either of the two laboratory manuals (Association of Official Agricultural Chemists 1970; Ministry of Agriculture, Fisheries, and Food 1973) listed at the end of this section.

In addition to the over-all amounts of different elements, considerable interest surrounds the levels of certain important anions. For example, the level of nitrate is significant from the point of view of the crop physiologist with an interest in plant development and protein synthesis in the whole plant. Nitrate levels are also important in relation to the feeding of ruminant animals which are poisoned by excessive amounts. Nitrate is most conveniently determined with an ion-specific electrode (section 4.4.3). This technique is also applicable to the determination of chloride ion. An alternate procedure for the latter ion is based on the precipitation of chloride with silver nitrate, followed by titration of the excess silver remaining in solution.

5.9. Compounds of special nutritional significance

It is mainly the bulk components of plant materials that are of primary importance in determining nutritive value. However, certain compounds which occur in relatively small amounts can also exert a considerable influence on the nutritional significance of a crop. These compounds include a range of toxic factors of great chemical diversity, together with the various vitamins which occur in plants.

TABLE 5.10

Some plant materials containing toxic compounds

Material	Compound(s)†	Significance
Soya beans	Polypeptides of molecular weight 16 000–24 000	Inhibitors of proteolytic enzymes
Castor beans	Ricin: a protein of molecular weight 85 000	Causes agglutination of red blood cells
Oilseed rape	Thioglucosides	Anti-thyroid compounds
Cassava	Linamarin: a cyanogenetic glucoside	May breakdown by autohydrolysis to yield hydrogen cyanide
Cottonseed	Gossypol pigments	Cause a wide range of deleterious symptoms in non-ruminant animals
Grass pea (*Lathyrus sativus*)	Neurotoxic amino acids	Can induce neurological disorders in man
Bracken	Thiaminase	May induce thiamine deficiency
Ragwort	Pyrrolizidine alkaloids	Can lead to liver damage and death of cattle
Badly preserved crops	Mycotoxins	Induce a range of adverse symptoms

†The toxic principles listed in this table have been widely studied and in most cases their chemical nature has been elucidated. However some of the plants mentioned also contain other harmful substances. Soya beans for example, also contain a goitrogenic compound and a potent lipoxygenase. Bracken contains a carcinogen in addition to the thiaminase and some workers consider the former substance, although less studied, to be of greater significance than the antivitamin.

5.9.1 *Toxic compounds*

Toxic compounds may occur in crop materials as natural constituents, as compounds of microbiological origin, or as components of contaminating plant tissues. Some selected examples are listed in Table 5.10. They include compounds of high molecular weight, such as the polypeptide *ricin*, which occurs in castor beans, and smaller compounds such as the *thioglucosides*, found in oilseed rape. As well as this diversity in molecular size and nature, there are also differences in the origins of the various toxic principles. Some occur as *contaminants* in an otherwise safe feedstuff. Examples include the *mycotoxins*, which are synthesized by fungi growing on poorly preserved hay or other materials, and the *Senecio* alkaloids which may poison livestock feeding on hay or silage containing large quantities of ragwort. In the United Kingdom *S. jacobaea* and *S. aquaticus* are common weeds which both contain alkaloids whose structure is based on the *pyrrolizidine* ring. The former species contains five alkaloids, including *Senecionine*.

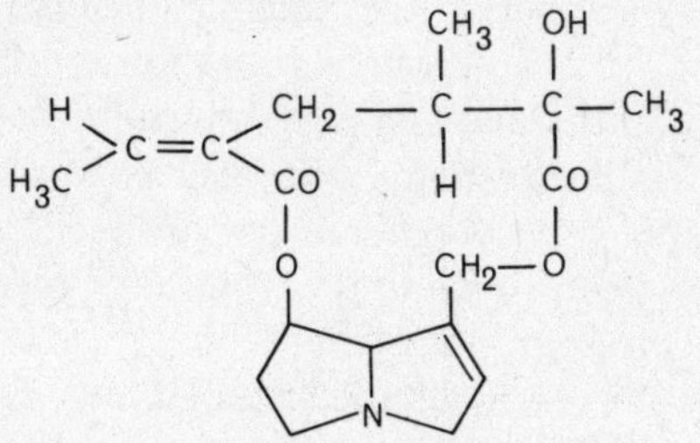

Senecionine

The other four alkaloids all contain the same heterocyclic double ring system but differ in their substituent groups. Although livestock tend to avoid the weed during grazing, this selectivity does not occur when preserved herbage is fed, and cases of multiple deaths within single herds are well documented.

Plant materials which contain toxic compounds as natural constituents include seeds, leaves, and roots. A number of seeds contain protease inhibitors, among them soya bean (*Glycine max*), lima bean (*Phaseolus lunatus*), and some other legumes. These inhibitors are proteins which can be denatured by heat treatment and rendered non-toxic. Soya bean meal, which contains a trypsin inhibitor, is routinely subjected to heat processing prior to use in animal feedstuffs. Many other seeds contain toxins, including those of the cotton plant (*Gossypium* spp.) and the grass pea (*Lathyrus sativus*).

Amongst root crops which contain toxic factors is the tapioca plant, or cassava (*Manihot utilissima*). The toxic principle in this case is *linamarin*, a member of a group of compounds known as *cyanogens*. These are glycosides which upon hydrolysis yield hydrogen cyanide.

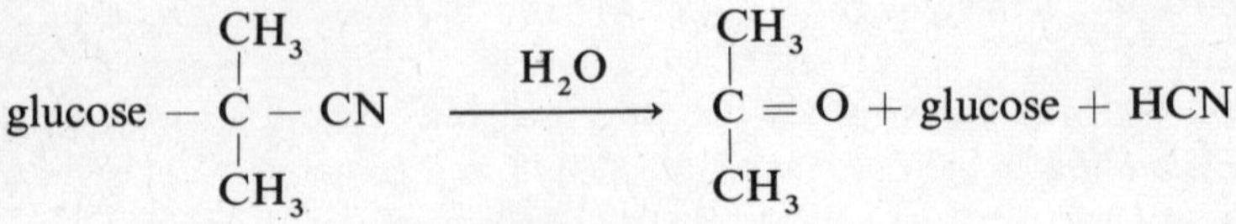

Linamarin

A well-known example of the occurrence of harmful compounds in leaves is the presence of an antithiamine factor in bracken (*Pteridium aquilinum*). Although this species is of little value as a forage, it is widespread in many areas of rough grazing. The toxic factor leads to a breakdown of thiamine, which can result in a vitamin deficiency in animals consuming the plant in large amounts.

Because of their great diversity with respect to molecular structure, reactivity, and physico-chemical nature, it is difficult to formulate any general approach to the analysis of toxic compounds. In a recent detailed monograph, Liener (1969) reviews many suitable methods, and recourse to a detailed text of this type is essential before studies are undertaken. As a single example, the analysis of the ragwort alkaloids will be considered. The common ragwort, *Senecio jacobaea*, contains a number of alkaloids, all of a very similar structure. Extraction can be achieved by the Soxhlet procedure using methanol as a solvent. Fresh material is placed directly in the thimble and refluxing is continued for several hours. Thin layer chromatography is suitable as a means of obtaining a semi-quantitative estimate of alkaloid content. The methanol extract is evaporated to dryness and the alkaloids re-dissolved in dilute hydrochloric acid. The solution is then freed of pigments by partition against ether. After the addition of ammonia the alkaloids are extracted into chloroform and taken for t.l.c. Separations are carried out on alkaline silica gel, using methanol as a developing solvent. Detection is with Dragendorff's reagent applied as a spray (Bull, Culvenor, and Dick 1968). The major alkaloids give an orange colour, at an R_f value of $\sim$0·4. The intensity of the colour and the degree of concentration of the extract required to give a positive analysis, gives some idea of the level of alkaloid in the sample. For more accurate quantitative work a spectrophotometric method is available (Mattocks 1967).

The essential steps in alkaloid analysis are extraction, purification, separation, and quantitative assay. These steps are also followed in the assay of many other toxic compounds. Occasionally much simpler procedures are available. For example, the level of cyanogens can be estimated by hydrolysis followed by assay of the liberated hydrogen cyanide. Some compounds are best examined by direct measurement of their inhibitory effect on enzymes, or other systems, while others can be estimated only after a series of fractionation steps.

5.9.2. *Vitamins*

Vitamins are required in small amounts for normal growth in animals. They are usually classified as either *fat-soluble* or *water-soluble* compounds. Many crops and their products are important sources of a wide range of vitamins and play an important role in maintaining dietary intakes at suitable levels. The importance of some different crops in relation to vitamin content is outlined in Table 5.11. The vitamins A, D, E, and K are fat soluble. Vitamin A does not exist in an active form in plants, but large quantities of its precursors or *provitamins* are found, particularly in green crops. The most abundant of these compounds is β-*carotene*. Vitamin D is not usually found in plants, except in roughages which have been dried in sunlight, where photochemically induced synthesis from naturally occurring sterols may take place.

The water soluble vitamins include vitamin C, together with members of the B-group complex. Many of the B vitamins are fairly widely distributed although some plant materials are particularly good sources of certain individual members of this group.

The estimation of vitamins present in crop materials may involve either a chemical procedure with the appropriate preliminary separation steps, or a bioassay using laboratory animals or microorganisms (Association of Official Agricultural Chemists 1970). Many of the chemical or physico-chemical methods are based on spectrophotometric or fluorimetric analysis. For example, the estimation of riboflavin is carried out by measuring fluorescence at 520 nm

TABLE 5.11
Some important sources of vitamins and provitamins

Crop	Vitamins of the B group	Other vitamins and provitamins
Cereal grains	Thiamine	Vitamin E
	Vitamin B_6	
	Biotin	
	Choline	
	Pantothenic acid	
Peas, beans, and other large grains	Thiamine	
	Nicotinamide	
	Vitamin B_6	
	Pantothenic acid	
Leafy crops	Riboflavin	Vitamin C
	Biotin	Vitamin E
	Choline	Vitamin K
	Thiamine	Carotenoids
Citrus fruits		Vitamin C
Carrots		Carotenoids
Potatoes		Vitamin C

following irradiation at 428 nm, while the level of carotene can be measured by determining the absorbance of a solution at 450 nm. In most cases, prior fractionation procedures are essential to avoid the presence of interfering compounds. Sometimes, a bioassay method and a chemical method are equally suitable alternatives and the choice then often depends on the laboratory facilities which are available.

5.10 Miscellaneous water-soluble constituents

Many water-soluble compounds occur in trace amounts as intermediaries in metabolic pathways. Despite their considerable importance in this respect they are of relatively little significance in determining the gross composition or properties of crops or their products. However, certain other water-soluble constituents may accumulate in plant tissues to fairly high levels and it is these compounds which will be discussed in this Section. Water-soluble constituents which have been discussed in the earlier part of this text will not be re-examined.

5.10.1 *Anthocyanins and betacyanins*

The *anthocyanins* are a group of pigments widely distributed in higher plants. They are found especially in flower parts and also in leaves, stems and fruits. They are coloured either pink, red, violet, or blue, according to chemical structure. Almost all plant tissues exhibiting these colours contain large amounts of one or more anthocyanins. One familiar exception is the beetroot which contains the glycoside, *betanin*. This nitrogenous pigment belongs to the *betacyanin* group.

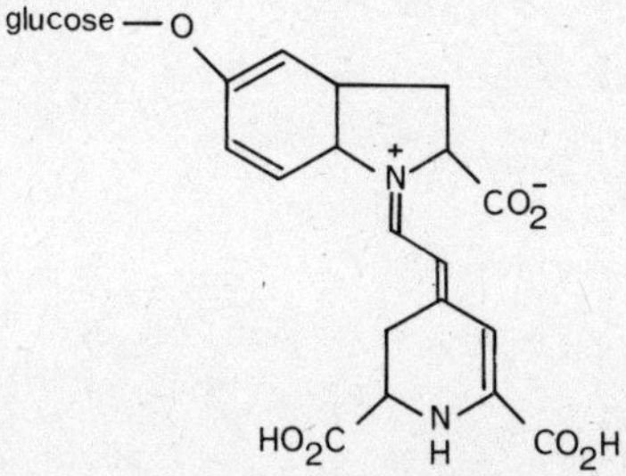

Betanin

The anthocyanins are also glycosides. One common example is *cyanidin 3-glucoside*.

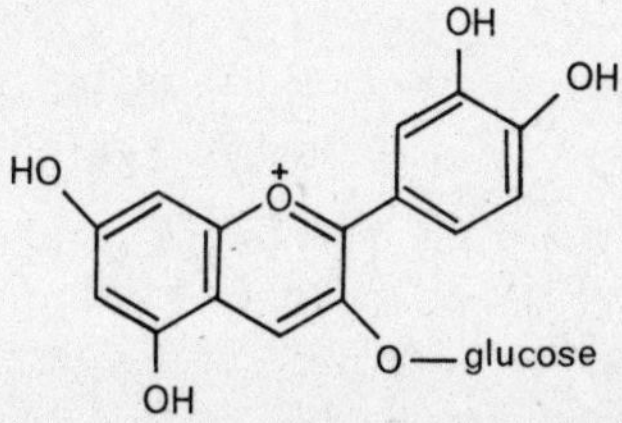

Cyanidin 3-glucoside

The glucose molecule can be removed by hydrolysis to give the corresponding aglycone, or *anthocyanidin*.

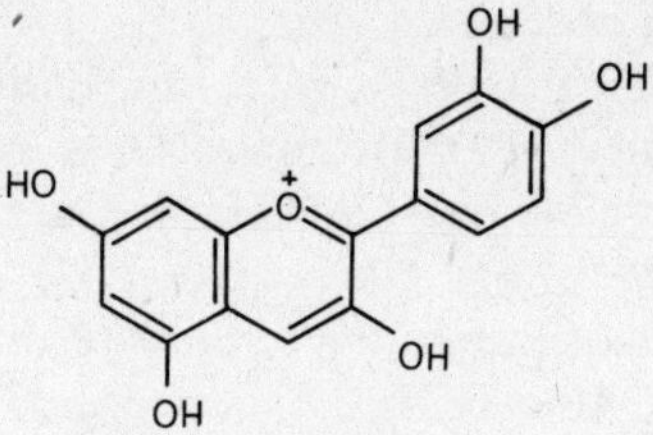

Cyanidin

BANGOR
LIBRARY

All the anthocyanidins contain the *flavylium ion* as an integral part of their structure.

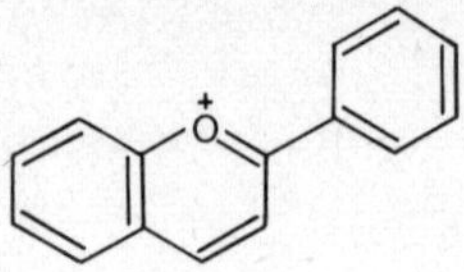

Flavylium ion

As well as cyanidin, there are five other common anthocyanidins. These include *pelargonidin* and *delphinidin*.

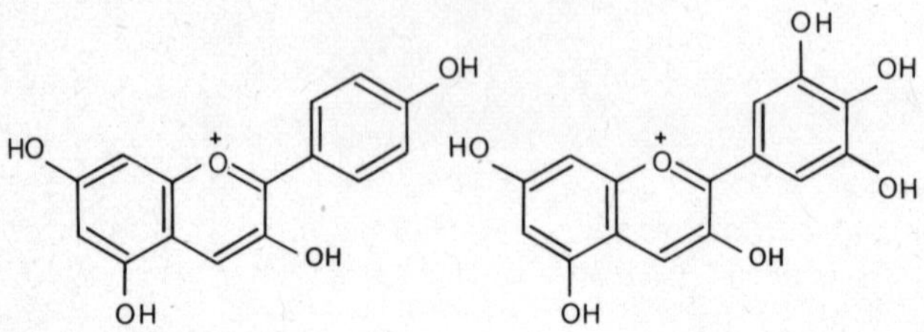

Pelargonidin Delphinidin

Variations in the nature and number of sugar residues substituted in these structures give rise to a range of different anthocyanins, all of which are water soluble. Although tissues can, in fact, be extracted with water as a first step in the analysis of these pigments, a more suitable solvent is methanol containing a small quantity of hydrochloric acid (1 ml conc. HCl per 100 ml methanol) (Swain 1965).

The anthocyanidins can be obtained by boiling fresh tissue with dilute HCl (2M). After this hydrolysis treatment flavones are removed with ethyl acetate and the anthocyanidins recovered from the aqueous phase by extraction into amyl alcohol. This procedure is due to Harborne (1973). Review articles containing details of chromatographic procedures applicable to the anthocyanins include those of Swain (1965) and Harborne (1967). Separations can be carried out on paper or thin layers of silica gel or cellulose. Quantitative analyses can be achieved by the spectrophotometric examination of solutions prepared with ethanol, or methanol, containing a small quantity of hydrochloric acid.

5.10.2 *Organic acids*

An unusual feature of many plant cells is their ability to accumulate quite large amounts of various organic acids in the cell vacuole. In grasses the level of total organic acids may be as high as 6 g per 100 g dry matter. The presence of such compounds in herbage is of some practical significance in relation to the *ensilage* process. The preservation of green crops as silage involves the fermentation of compounds present in the cells of the leaf tissues. *Organic acids* are fermented by heterolactic bacteria which are present naturally in the harvested crop. The products include lactic acid, which acts as a preservative during winter storage of the fermented material, and carbon dioxide, which is lost to the atmosphere. This loss of material is a small, but significant part of the over-all dry matter wastage during the silage-making process.

Organic acids are also important constituents of many fruits. Their presence in large amounts results in a characteristic 'sharp' taste. The quantitative analysis of fruit acidity by titration has, in fact, been used in the practical evaluation of fruit quality.

Although one particular acid may predominate in a particular species it is usual to find a wide range of acidic components present within a single tissue. Thus, although the main acid in the apple is *malic*, twenty others have also been identified. These include another Krebs-cycle intermediate, *citric acid*, together with *shikimic acid*, an important intermediate in the synthesis of the aromatic amino acids.

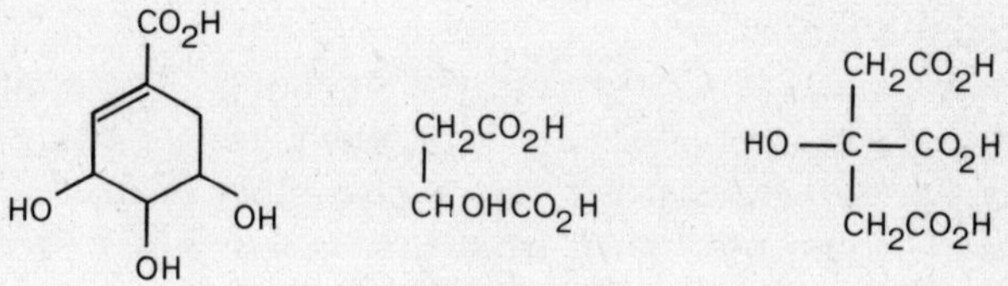

Shikimic acid Malic acid Citric acid

These three examples illustrate another interesting aspect of the chemical diversity of plant acids, namely the occurrence of varying numbers of carboxylic acid groups. Thus, with shikimic, malic, and citric acid we have examples of *mono-*, *di-*, and *tricarboxylic* acids respectively.

Although of rather narrow distribution, *toxic* organic acids are known to occur in appreciable amounts in certain plants. In the leaves of the rhubarb plant (*Rheum rhaponticum*) there is an accumulation of *oxalic* acid which has occasionally led to cases of poisoning. Another toxic compound, which occurs elsewhere, is *monofluoracetic acid*.

The first step in the analysis of organic acids is to prepare an extract with water or boiling ethanol. Purification of the extract can be achieved by ion-exchange chromatography and individual acids isolated by paper chromatography (Harborne 1973). Separations can also be carried out by g.l.c. following a preliminary methylation step (Mazliak and Salsac 1965).

5.11. Exogenous materials

Exogenous materials which occur in crops and their products may include herbicides, fungicides, and other pesticidal compounds. Contamination may also occur as a result of the discharge of industrial effluents, either into watercourses or directly onto the land. In recent years, an increased awareness of the possible accumulation or persistence of toxic compounds in the environment has stimulated research into the fate of pesticides within plant tissues, and also into the uptake and distribution of other exogenous materials. Studies of this type have frequently provoked a lively debate regarding the advisability of the continued use of certain agrochemicals. It is almost certain that this discussion will continue, as often scientists presented with the same data arrive at different conclusions. An example which illustrates this point occurs in relation to the use of mercurial fungicides.

In a recent article Calderbank (1974) states: 'The quantities of mercury used in agriculture are very much smaller than the waste arising from industrial usage, so the possibility of affecting fish or causing environmental harm from this source *can be discounted*.' Whereas Finlayson and MacCarthy (1973) conclude: 'Internal residues of mercury in plants are not yet a problem, but it is clear that *the use of mercury in fungicides must be reduced or ended*. It is now realised that mercury can be converted into an organic and more deadly form and then concentrated by microorganisms.'

It would be inappropriate in the present text to pursue this discussion, other than to comment that a continued interest in the analysis of toxic residues will remain an essential part of the scientist's role in agriculture, particularly when intensive and highly mechanized systems of husbandry are practised.

5.11.1 *Pesticides and herbicides*

The safety evaluation of pesticides in relation to their persistence in the environment is complicated by the fact that often the original compound is rapidly broken down, giving a range of products which may or may not retain the toxic properties of the applied substance. For example, *malathion*, an *organophosphorous* pesticide, gives rise to a whole series of compounds which must then be studied individually (Fig. 5.11).

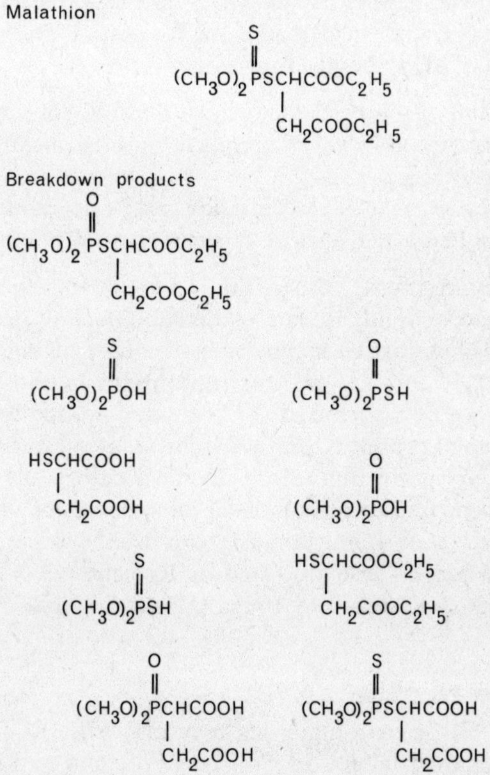

FIG. 5.11. Malathion and some products formed during degradation.

Other difficulties also arise, mainly as a result of the fact that the levels of pesticides encountered in plant tissues are usually extremely low. Thus the analytical procedure which is chosen must be highly selective. Interference may arise, not only from naturally occurring plant products, but also from other synthetic materials which have become widespread in the biosphere. An example occurs with the *organochlorine pesticides*. Their analysis is subject to interference by the *polychlorinated biphenyls* which are used in the manufacturing industries and have, as a consequence, become widely distributed in nature. Interference may also occur as a result of the contamination of solvents or reagents used in the assay with compounds which give a positive response in the analytical system. These problems can be overcome by obtaining a positive identification of the residue after the final separation.

In a review of analytical procedures for pesticide residues, Ruzicka (1973) divides analysis into four stages:

(a) extraction of the pesticide using a *selective* solvent,
(b) chromatography or solvent partition as a means of *removing interfering substances*,
(c) *quantitative assay* of the pesticide and its degradation products,
(d) *confirmation of the identity* of the compounds assayed.

The chromatographic procedure adopted for the removal of interfering compounds is usually a column technique involving the use of *Florisil* (activated magnesium silicate) packing material. The eluate is then assayed by an appropriate method. Gas–liquid chromatography is preferred for this purpose due to its high sensitivity. The conditions required for the assay of different pesticides obviously vary according to the chemical nature of the compound and also according to the presence, or absence, of interfering substances. It is thus important to refer to the original reports of particular assays. These are listed by Ruzicka (1973). The analysis of herbicides is discussed by Freed (1964).

5.11.2 *Other exogenous materials*

The pollution of crop materials may also arise as a result of the disposal of industrial wastes. Sources of contamination are gaseous or particulate discharges into the atmosphere, liquid effluents released into rivers or streams, and solid or semi-solid materials dumped in quarries or on waste-land. The polluting substances are frequently heavy metal elements which may possess highly toxic characteristics. The detection of such contamination can usually be achieved by standard procedures such as spectrophotometry or atomic absorption or emission-analysis (section 4.1). Fortunately, many metals are poorly absorbed by plants, especially when the element in question is present in the soil in a non-ionic form. Although many of the methods used in this type of work are fairly straightforward, it is important to take into account the possible localized accumulation of residues in specific parts of the plant, and in the evaluation of the practical significance of results, to consider the status of the crop product in the diet of man or domestic animals.

REFERENCES

Books and review articles

ALTSCHUL, A. M., YATES, L. Y., ORY, R. L., and ENGLEMAN, E. M. (1966). Seed Proteins. *A. Rev. Pl. Physiol.* **17**, 113.

ASPINALL, G. O. (1970). *Polysaccharides.* Pergamon Press, Oxford.

ASSOCIATION OF OFFICIAL AGRICULTURAL CHEMISTS (1970). Methods of analysis. A.O.A.C., Washington.

AUDUS, L. J. (1972). *Plant growth substances.* Leonard Hill, London.

BARNES, R. F. (1973). Laboratory methods of evaluating feeding value of herbage. In *Chemistry and biochemistry of Herbage*, Vol. 3 (ed. G. W. Bulter, and R. W. Bailey), p. 179. Academic Press, London and New York.

BULL, L. B., CULVENOR, C. C. J., and DICK, A. T. (1968). *The pyrrolizidine alkaloids.* North Holland Publishing Company, Amsterdam.

BURCHFIELD, H. P., and STORRS, E. E. (1962). *Biochemical applications of gas chromatography.* Academic Press, New York and London.

CALDERBANK, A. (1974). The use of herbicides and fungicides. In *Pollution and the use of chemicals in agriculture* (ed. D. E. G. Irvine and B. Knights), p. 26. Butterworths, London.

CRIDDLE, R. S. (1966). Protein and lipoprotein organization in the chloroplast. In *Biochemistry of Chloroplasts* (ed. T. W. Goodwin), p. 203. Academic Press, London and New York.

DIXON, M., and WEBB, E. C. (1964). *Enzymes* (2nd ed). Longmans, New York.

FINLAYSON, D. G., and MACCARTHY, H. R. (1973). Pesticide residues in plants. In *Environmental pollution by pesticides* (ed. C. A. Edwards), p. 57. Plenum Press, London and New York.

FREED, V. H. (1964). Determination of herbicides and plant growth regulators. In *The physiology and biochemistry of herbicides* (ed. L. J. Audus). Academic Press, London and New York.

HARBORNE, J. B. (1967). Chromatography of phenolic compounds. In *Chromatography* (ed. E. Heftmann), (2nd edn), p. 677. Reinhold Publishing Co., New York.

—— (1973). *Phytochemical methods*, p. 33. Chapman and Hall, London.

HARKIN, J. M. (1973). Lignin. In *Chemistry and biochemistry of herbage*, Vol. 1 (ed. G. W. Butler and R. W. Bailey), p. 323. Academic Press, London and New York.

HITCHCOCK, C., NICHOLS, B. W. (1971). *Plant lipid biochemistry.* Academic Press, London and New York.

HOLLIGAN, P. M. (1971). A Review of techniques used for gas–liquid chromatography of carbohydrates. *New Phytol.* **70**, 239.

INTERNATIONAL UNION OF BIOCHEMISTRY (1961). *Report of the Commission on Enzymes.* Pergamon Press, London.

LIENER, I. E. (1969). *Toxic constituents of plant foodstuffs.* Academic Press, London and New York.

McDONALD, P., EDWARDS, R. A., and GREENHALGH, J. F. D. (1973). *Animal nutrition.* Oliver and Boyd, Edinburgh.
MARTIN, J. T., and JUNIPER, B. E. (1970). *The cuticles of plants.* Arnold, London.
MEISTER, A. (1957). *Biochemistry of the amino acids.* Academic Press, New York.
MINISTRY OF AGRICULTURE, FISHERIES AND FOOD (1973). *The analysis of agricultural materials.* H.M.S.O., London.
OSBORNE, T. B. (1909). *The vegetable proteins.* Longmans Green & Co., London.
ROTHFIELD, L. and FINKELSTEIN, A. (1968). Membrane biochemistry. *A. Rev. Biochem.* **37**, 463.
STAHL, E. (1965). *Thin layer chromatography.* Springer-Verlag, New York.
SWAIN T. (1965). Analytical methods for flavonoids. In *Chemistry and biochemistry of plant pigments* (ed. T. W. Goodwin), p. 533. Academic Press, London and New York.
RUZICKA, J. H. (1973). Methods and problems in Analysing for Pesticide Residues in the environment. In *Environmental pollution by pesticides* (ed. C. A. Edwards), p. 11. Plenum Press, London and New York.
WEAVER, R. J. (1972). *Plant growth substances in agriculture.* Freeman, San Francisco.
WHELAN, W. J. (1955). Starch, glycogen, fructosans and similar polysaccharides. In *Modern methods of plant analysis* (ed. K. Paech and M. V. Tracey), Vol. II, p. 145. Springer-Verlag, Berlin.

Research papers

ANDERSON, J. W. (1968). Extraction of enzymes and subcellular organelles from plant tissues *Phytochemistry* **7**, 1973.
CAVELL, B. D., MACMILLAN, J., PRYCE, R. J., SHEPPARD, A. C. (1967). Thin layer and gas–liquid chromatography of the gibberellins; direct identification of the gibberellins in a crude plant extract by gas–liquid chromatography. *Phytochemistry* **6**, 867.
CLANCY, M. J. and WILSON, R. K. (1966). Development and application of a new chemical method for predicting the digestibility and intake of herbage samples. *Proc. int. Grassal Congr.* **12**. 95.
CRAMPTON, E. W.and MAYNARD, L. A. (1938). The relation of cellulose and lignin content to the nutritive value of animal feeds. *J. Nutr.* **15**, 383.
CRIDDLE, R. S. (1969). Structural proteins of chloroplasts and mitochondria. *A. Rev. Pl. Physiol.* **20**, 239.
DAVIS, L. A., HEINZ, D. E., and ADDICOTT, F. T. (1968). Gas-liquid chromatography of trimethylsily derivatives of abscisic acid and other plant hormones. *Pl. Physiol., Lancaster* **43**, 1389.
DETCHOU, P. and POSSINGHAM, J. V. (1972). Ribosomal-RNA distribution during leaf development in spinach. *Phytochemistry* **11**, 943.
DRAPER, S. R. (1973). Amino acid profiles of chemical and anatomical fractions of oat grains. *J. Sci. Fd Agric.* **24**, 1241.
DUNCOMBE, W. G. (1963). The colorimetric micro-determination of long-chain fatty acids. *Biochem. J.* **88**, 17.
EAGLES, C. F. and WAREING, P. F. (1964). The role of growth substances in the regulation of bud dormancy. *Physiologia Pl.* **17**, 697.
ELLIS, G. H., MATRONE, G., and MAYNARD, L. A. (1946). A 72 per cent H_2SO_4 method for the determination of lignin and its use in animal nutrition studies. **J.** *Anim. Sci.* **5**, 285.
FOLKES, B. F. and YEMM, E. W. (1956). The amino acid content of the proteins of barley grains. *Biochem. J.* **62**, 4.

GALLIARD, T., RHODES, M. J. C., WOOLTORTON, L. S. C., and HULME, A. C. (1968). Metabolic changes in excised fruit tissue—III. The development of ethylene biosynthesis during the ageing of apple peel. *Phytochemistry* **7**, 1465.
GILES, K. W. and MYERS, A. (1966). The effects of gibberellic acid and light on RNA, DNA and growth of the three basal internodes of dwarf and tall peas. *Phytochemistry* **5**, 193.
HAMILTON, R. J. and POWER, D. M. (1969). The chemical composition of the surface wax of *Lolium perenne*. *Phytochemistry* **8**, 1771.
MACRAE, J. C. and ARMSTRONG, D. G. (1968). Enzyme method for determination of α-linked glucose polymers in biological materials. *J. Sci. Fd Agric.* **19**, 578.
MACKINNEY, G. (1941). Absorption of light by chlorophyll solutions. *J. biol. Chem.* **140**, 315.
MACEY, M.J.K. and BARBER, H.N. (1970). Chemical genetics of wax formation on leaves of *Brassica oleracea*. *Phytochemistry* **9**, 13.
MATTOCKS, A. R. (1967). Spectrophotometric determination of unsaturated pyrrolizidine alkaloids. *Analyt. Chem.* **39**, 443.
MAZLIAK, P. and SALSAC, L. (1965). Analyse, par chromatographie en phase gazeuse, des acides organiques vegetaux. *Phytochemistry* **4**, 693.
MILLBORROW, B. V. (1967). The identification of (+)-abscisin II ((+)-dormin) in plants and measurement of its concentrations. *Planta* **76**, 93.
MILNER, C. K. and WESTGARTH, D. R. (1973). A collaborative study of the determination of available lysine in feeding stuffs. *J. Sci. Fd Agric.* **24**, 873.
MOORE, S. and STEIN, W. H. (1954). A modified ninhydrin reagent for the photometric determination of amino acids and related compounds. *J. biol. Chem.* **211**, 907.
MORRISON, I. M. (1972). Improvements in the acetyl bromide technique to determine lignin and digestibility and its application to legumes. *J. Sci. Fd Agric.* **23**, 1463.
OSBORNE, D. J. and MCCALLA, D. R. (1961). Rapid bioassay for kinetin and kinins using senescing leaf tissue. *Pl. Physiol.* **36**, 219.
ROUGHAN, P. G. and BATT, R. D. (1968). Quantitative analysis of sulfolipid (sulfoquinvosyl diglyceride) and galactolipids (monogalactosyl and digalactosyl diglycerides) in plant tissues. *Analyt. Biochem.* **22**, 74.
ROUSER, G., SIAKATOS, A. N., and FLEISCHER, S. (1966). Quantitative analysis of phospholipids by thin layer chromatography and phosphorous analysis of spots. *Lipids* **1**, 85.
SOMOGYI, M. (1945). A new reagent for the determination of sugars. *J. biol. Chem.* **160**, 161.
UPPER, C. D., HELGESON, J. P., KEMP, J. D., and SCHMIDT, C. J. (1970). Gas–liquid chromatographic isolation of cytokinins from natural sources. *Pl. Physiol.* **45**, 543.
VAN STADEN, J. (1973). Changes in endogenous cytokinin levels during abscission and senescence of *Streptocarpus* leaves. *J. exp. Bot.* **24**, 667.
WAITE, R. and GORROD, A. R. N. (1959). The structural carbohydrates of grasses. *J. Sci. Fd Agric.* **10**, 308.
WAITE, R., JOHNSTON, M. J., and ARMSTRONG, D. G. (1964). The evaluation of artificially dried grass as a source of energy for sheep. *J. agric. Sci.* **62**, 391.
WEIHE, H. D. and PHILLIPS, M. (1947). The quantitative estimation of hemicelluloses by direct isolation. *J. agric. Res.* **74**, 77.
WEST, C. A. and REILLY, T. (1961). Properties of gibberellins from flowering plants. *Adv. Chem. Ser.* **28**, 37.

6. Data handling

THE advanced instrumentation and automation characteristic of many modern laboratories often results in an extremely rapid output of data. The processing of this information can be a time-consuming task unless appropriate measures are taken to minimize the number of operations requiring the attention of the experimenter. Fortunately, equipment is now available to deal with the large number of results which must be calculated each day. The output from gas chromatographs, automated analytical equipment, and optical instruments is usually in the form of a printed paper strip, or a continuous trace on a recorder chart, with each maximum corresponding to a particular component. In the latter case, peak heights can be determined with a chart reader, an electronic device which automatically measures the height of a peak above the base line after the operator has aligned a cursor over the trace. Subsequent steps in the processing of data can often be accomplished with one of the programmable calculators which are now available. Calculations can be further speeded up by the direct linkage of the chart reader to the calculator. This system is capable of giving the final answer as rapidly as the chart reader can be operated. Similar direct links can be made to computers, which, of course, are able to handle more complicated procedures than are possible with the simpler laboratory equipment. These techniques are almost essential in laboratories handling large numbers of samples for routine purposes. However, in many laboratories problems of data handling relate to single experiments rather than to routine procedures carried out day by day. The evaluation of data obtained in such experiments usually involves statistical analysis.

It is outside the scope of this book to discuss the use of statistical methods in the evaluation of experimental results; however, there are many other texts which deal with the application of these techniques to problems in biology. An excellent introduction is provided by Bishop (1971), who describes a number of the most useful and widely used tests and also discusses other practical aspects of the subject such as the planning of experiments and the use of calculating machines. For a more advanced treatment, reference should be made to the work of Snedecor (1956). The statistical tables required in the execution of these analyses are available as a separate publication (Fisher and Yates 1963). Many of the standard statistical analyses can be carried out with a programmable calculator of the type used in the computation of analytical results. In the absence of this facility, use can be made of less complex electronic calculators, providing adequate storage capacity is available. The applications of different types of calculators are summarized in Table 6.1.

TABLE 6.1

Electronic aids to data handling in the laboratory

Type of instrument	Applications
Fully portable pocket-size calculators	Replacement for the slide rule giving greater accuracy and speed of use
Bench-top calculators with storage capacity and facilities for squares and square-roots	Calculations involving the use of data storage capacity, including statistical analysis and the computation of experimental results
Programmable calculators	Rapid statistical analysis of data and fully automated calculation of the results of routine assays

REFERENCES

Books and review articles

BISHOP, O. N. (1971). *Statistics for Biology* (2nd edn). Longman. London.

FISHER, R. A. and YATES, F. (1963). *Statistical tables for biological, agricultural and medical research* (6th edn). Oliver and Boyd, Edinburgh.

SNEDECOR, G. W. (1956). *Statistical methods.* Iowa State University Press.

Appendix: The composition of some crops and their products

	Dry matter (g per 100g fresh material)	Crude protein	Total lipid	Crude fibre	Ash
		(g per 100 g dry mater)			
Fresh leafy crops					
Pasture grass (leafy)	19	17·4	2·6	23·7	11·6
Pasture grass (flowering)	23	10·4	2·2	27·0	11·6
White clover	19	23·1	4·2	22·6	11·1
Cabbage	11	13·6	3·6	18·2	10·9
Kale	14	15·7	3·6	17·9	3·6
Alfalfa (early flowering)	24	17·1	1·7	30·0	10·0
Preserved leafy crops					
Hay	85	7·0	1·9	35·5	7·1
Grass silage	20	15·2	4·8	21·5	10·0
Dried grass	90	16·7	2·4	23·2	12·0
Barley straw	86	3·5	2·1	39·3	5·3
Roots and tubers					
Swedes	12	9·3	0·7	12·0	7·4
Turnips	9	12·0	2·2	10·0	7·8
Potatoes	24	8·8	0·4	3·8	4·2
Carrots	13	9·2	1·5	10·8	6·9
Sugar beet	23	4·8	4·4	4·8	3·1
Mangolds	12	8·3	0·8	5·8	6·7
Jerusalem artichoke	20	7·4	1·0	3·4	5·4
Parsnips	15	8·7	2·0	8·0	6·0
Cassava	32	3·6	1·0	4·6	3·9
Sweet-potato	31	5·4	1·3	4·2	3·6
Yam (*Dioscorea alata*)	69	2·9	0·2	1·5	2·0
Cereal grains					
Barley	85	10·3	1·8	5·3	3·1
Oats	87	12·0	5·5	11·9	3·6
Wheat	87	14·0	2·2	2·2	2·0
Maize	87	11·4	5·1	2·5	1·5
Rice	89	9·3	2·0	9·9	5·6
Rye	87	13·2	1·9	2·2	2·3
Sorghum (*S. vulgare*)	89	12·2	3·4	2·7	2·1
Millet (foxtail)	89	13·5	4·6	9·3	4·0
Millet (proso)	90	12·8	4·0	7·2	3·0

	Dry matter (g per 100g fresh material)	Crude protein	Total lipid	Crude fibre	Ash
		(g per 100g dry matter)			
Other seeds					
Peas	86	26·1	1·9	6·3	3·3
Field bean	86	29·7	1·8	8·3	3·7
Soya bean	90	36·9	19·4	4·6	5·2
Rape seed	93	21·2	48·6	6·4	4·5
Groundnut	94	28·4	47·6	2·8	2·3
Sunflower	93	15·3	34·9	30·4	3·7
Locust bean (including pod)	85	6·8	1·5	7·6	3·0
Lentils	86	29·6	2·2	3·9	3·5
Mung bean	90	26·6	1·4	4·3	4·2
Lima bean	90	23·1	1·5	5·1	4·6
Chick pea	89	21·9	4·4	7·9	3·3
Fruit crops					
Orange	16	7·5	1·9	11·3	4·4
Grapefruit	14	8·1	4·4	10·3	3·7
Tomato	6	16·4	5·0	9·1	8·2
Pumpkin	9	16·2	8·9	14·2	8·9
Banana	24	4·5	0·8	2·1	3·3

Notes: 1. Total lipid estimates are obtained by ether extraction.

2. Sources of reference:
   - (a) McDonald, P., Edwards, R. A., and Greenhalgh, J. F. D. (1973). *Animal nutrition* (2nd edn). Oliver and Boyd, Edinburgh.
   - (b) Evans, R. E. (1960). *Rations for livestock*. Ministry of Agriculture, Fisheries and Food Bulletin, no. 48. H.M.S.O., London.
   - (c) National Academy of Sciences (1971). *Atlas of Nutritional Data on United States and Canadian Feeds*. Washington.
   - (d) Coursey, D. G. (1967). *Yams*. Longmans, London.

   Other values are average figures recorded by staff of the School of Agriculture, Abedeen.

3. The figures quoted are typical values to be expected for the crops in question. However, variations occur as a result of environmental and genetic factors and where there is a requirement for accurate information it is important that chemical analysis are carried out on individual samples.

Index

UNIV. COLL. N.W.
BANGOR
LIBRARY